AF378874

MODELING IN ANALOG DESIGN

MODELING IN ANALOG DESIGN

Edited by

Jean-Michel Bergé
France Telecom-CNET

Oz Levia
Synopsis

and

Jacques Rouillard
ESIM

KLUWER ACADEMIC PUBLISHERS
BOSTON / DORDRECHT / LONDON

A C.I.P. Catalogue record for this book is available from the Library of Congress

ISBN 0-7923-9569-7

Published by Kluwer Academic Publishers,
P.O. Box 17, 3300 AA Dordrecht, The Netherlands.

Kluwer Academic Publishers incorporates
the publishing programmes of
D. Reidel, Martinus Nijhoff, Dr W. Junk and MTP Press.

Sold and distributed in the U.S.A. and Canada
by Kluwer Academic Publishers,
101 Philip Drive, Norwell, MA 02061, U.S.A.

In all other countries, sold and distributed
by Kluwer Academic Publishers Group,
P.O. Box 322, 3300 AH Dordrecht, The Netherlands.

Printed on acid-free paper

SERIES PRESENTATION

Current Issues in Electronic Modeling is a series of volumes publishing high quality, peer-reviewed papers dealing with modeling issues in the electronic domain. The objective is to establish a unique communication channel between academia and industry which will serve the growing needs in the field of modeling.

Electronic Hardware modeling is the art of describing a device, or a system, for a given purpose (simulation, synthesis, etc.) by the use of preestablished conventions (charts, languages). Modeling is attractive since it provides for simplification and abstraction of the actual object or process. Using common modeling techniques and conventions facilitates communication and information sharing and helps speed the process of design, fabrication, testing and manufacturing.

Over the last several years there has been a dramatic increase in the development and use of modeling languages, models, and modeling techniques. The overall objective of this series is to provide a platform for dissemination of the basic concepts, techniques, and applications of modeling.

Needs: Although much literature on modeling is available, it is widely distributed. This dedicated series has been initiated for the following reasons:

• First, the interest in the discipline of modeling is growing rapidly as more and more individuals and organizations become involved in complex electronic design.

• Second, it is evident that modeling transcends many levels of the process of electronic design as well as multi-disciplines. This series brings together users and generators and provide a platform for sharing theory, practice and application of modeling, modeling methods, and modeling techniques.

Scope: The series covers, but is not limited to, the following topics:

• Modeling practice in the area of Simulation, Synthesis, Timing, Analog, and other domains.

• Languages and paradigm issues regarding languages, Objet oriented modeling, Mixed level modeling, and System level modeling.

• Meta-modeling, (i.e. modeling the process of modeling), and modeling models.

• Information modeling in the domain of electronic design.

• Specification, formal methods and languages, and the validation models.

We hope you will enjoy reading this series. We welcome your suggestions and look forward to having you as a contributor.

The Series Editors

Jean-Michel Bergé, France Telecom-CNET
Email: berge@cns.cnet.fr
Oz Levia, Synopsys Inc.
Email: ozl@aol.com
Jacques Rouillard, ESIM
Email: rouillard@acm.org

EDITORS

Series Editors

Jean-Michel Bergé - *France Telecom - CNET*
Oz Levia - *Synopsys*
Jacques Rouillard - *ESIM*

Principal Advisor to the Editors

Jim Armstrong - *Virginia Tech*

Advisory Board

Raul Camposano - *Synopsys Inc.*
Hilary Kahn - *University of Manchester*
Zain Navabi - *University of Tehran*
Wolfgang Nebel - *University of Oldenburg*
Alec Stanculescu - *Fintronic Inc.*
Alain Vachoux - *Swiss Federal Institute of Technology*
Ron Waxman - *University of Virginia*

Editorial Board

Przemyslaw Bakowski - *IRESTE*
Dave Barton - *Intermetrics Inc.*
Bill Billowich - *VHDL Technology Group*
Mark Brown - *Compass*
Steve Carlson - *Escalade*
Simon Curry - *Cadence*
Tedd Corman - *EPS*
Alain Fonkoua - *ESIM*
Andreas Hohl - *Siemens AG*
Michael Hohenbichler - *CEC*
Sabine Maerz - *Siemens AG*

Serge Maginot - *Leda S.A.*
Wolfgang Mueller - *CADLAB*
Adam Pawlak - *GMD*
Bora Prazic - *Alcatel*
Paul Scheidt - *Synopsys Inc.*
Jean-Pierre Schoellkopf - *SGS-Thomson*
Steve Schultz - *Texas Instruments*
Ken Scott - *Synopsys Inc.*
Venkat Venkataraman - *IBM*
Alex Zamfirescu - *Intergraph Electronics*
Roger Zinsner - *Speed S.A*

VOLUME 2: MODELING IN ANALOG DESIGN

Analog modeling is currently undergoing a profound mutation.

Analog behavioral descriptions are the new trend in analog modeling. However, analog behavioral modeling is not new and has in fact existed (and been used) for decades!

To understand the present situation, we have to look at the how the use of analog languages has evolved. Five or ten years ago, two kinds of analog engineers coexisted (and indeed still do coexist today in many cases). The first, and most numerous kind usually modeled analog circuits by building netlists of predefined components. The language (or format) they used to assemble these parts was usually straightforward enough. These "SPICE netlists", i.e. structural descriptions, were extensively simulated using analog simulators. For users of structural analog languages, the behavioral description was hidden within predefined models.

The second category of analog designers was the model builders. Being much more skilled in software, they described the behavior of the predefined models using classical programming languages to achieve highly optimized implementations.

Today, analog behavioral languages are available to all analog designers. Predefined models still exist and are being used, but now designers have the power of directly expressing the behavior of parts in his or her circuit, in addition to using predefined parts.

What is at stake? Undoubtedly there are advantages in this new category of language. In the short term, by favoring the top-down design and raising the level of description abstraction, this approach provides greater freedom of implementation and a higher degree of technology independence. Futhermore certain descriptions are made easier thanks to the possibility of being able to use either a behavioral or structural style.

In the longer term, analog synthesis (the automatic process inferring hardware from a behavioral description) and formal optimization (the ability to formally compute and optimize a set of equations) are targeted.

At the heart of this evolution are two main language standardizations: VHDL-A, the IEEE 1076.1 extension of VHDL which focuses more on mixed-mode analog/digital descriptions, and MHDL, focussing more on the microwave domain.

One of the first volumes of "Current Issues in Electronic Modeling" had therefore to be devoted to this domain.

Chapter 1 and 2 are related to VHDL-A. Chapter 1 shows the scope of VHDL-A and explains its rationale, while chapter 2 presents the practical aspects of modeling a MOS device using VHDL-A. Chapter 3 presents MHDL; introduces the main concepts of MHDL from the language point of view and provides some modeling examples.

The two other chapters of this volume deal with a particularly interesting application of analog behavioral languages: the modeling of thermal and electro-thermal aspects.

Thermal modeling is motivated by two main considerations. First of all, technologies are becoming more and more integrated and thermal aspects are therefore becoming ever more crucial. Secondly, highly accurate techniques and tools are now available.

Chapter 4 presents the general problem of electro-thermal interactions and introduces the methodology to be followed to describe electro-thermal models. Chapter 5 focuses on the thermo-electrical aspects for the description of power MOSFETs and bipolar transistors.

We hope this volume provides a good introduction to the current issues in analog modeling. Due to the highly evolving nature of this domain, we expect to devote other issues to this aspect in the future.

Jean-Michel Bergé, France Telecom-CNET
Co-editor of the series

CONTENTS

CONTRIBUTORS

Rachid Bouchakour
Telecom-Paris,
Département d'Electronique,
46 rue Barrault,
75634 Paris Cédex 13, France

Ernst Christen
Analogy, Inc.,
P.O. Box 1669,
Beaverton,
Oregon 97075, USA

Christophe Lallement
Telecom-Paris,
Département d'Electronique,
46 rue Barrault,
75634 Paris Cédex 13, France

S. Peter Liebmann
Compass Design Automation,
5457 Twin Knolls Road,
Columbia,
MD21045, USA

H. Alan Mantooth
Analogy, Inc.,
P.O. Box 1669,
Beaverton,
Oregon 97075, USA

Thierry Maurel
Telecom-Paris,
Département d'Electronique,
46 rue Barrault,
75634 Paris Cédex 13, France

David L. Rhodes
Army Research Laboratory
Electronics and
Power Sources Directorate,
AMSRL-EP-MA,
Fort Monmouth,
NJ 07703-5601, USA

C.-J. Richard Shi
Analogy, Inc.,
Simulation Technology Group,
Beaverton, OR 97008, USA

Alain Vachoux
Swiss Federal Institute of Technology
Integrated Systems Center,
EPFL-DE-LEG/C3i,
CH-1015 Lausanne, Switzerland

1

VHDL-A DESIGN OBJECTIVES AND RATIONALE[1]

C.-J. Richard Shi*, Alain Vachoux**

*Analogy, Inc.,Simulation Technology Group, Beaverton, USA

**Swiss Federal Institute of Technology, Integrated Systems Center
Lausanne, Switzerland

ABSTRACT

This article describes and analyzes the basis of VHDL-A language design: VHDL-A design objectives. Built from user requirements, VHDL-A design objectives specify a set of concrete goals and constraints that VHDL-A language design must meet. In this paper, we categorize design objectives into several groups: language scope, structure description mechanism, behavior description mechanism, and mixed-mode simulation support. We then justify each design objective, and describe various engineering trade-offs behind each design objective. In addition, the advantages and challenges of building an analog hardware description language on top of VHDL are also discussed.

[1] C.-J. Shi's work was supported in part by Rome Laboratory of United States Air Force under Contract F30602-93-C-0209/CSC292. Alain Vachoux's work was supported in part by the Swiss Federal Office for Education and Science (as a participation to the European ESIP ESPRIT project 8370). Both authors are members of the IEEE 1076.1 Working Group that is currently defining analog extensions to VHDL.

J.-M. Bergé et al. (eds.), Modeling in Analog Design, 1–30.
© 1995 Kluwer Academic Publishers. Printed in the Netherlands.

1.1. INTRODUCTION

Inspired by the success of VHDL as an industry standard of hardware description language in the design automation of digital electronic circuits, the VHDL-A initiative is taken to meet the ever increasing demand of the analog modeling capacity in VHDL from four application areas:

• **System design:** More than 80 percent of printed circuit board design, 60 percent of custom IC design and virtually all system-level electronic designs use a mix of analog and digital elements [1].

• **System verification:** Full system verification of an electronic design needs modeling of its non-electrical application environments (e.g. mechanical, thermal), which are analog in nature.

• **Digital design:** Cells and interconnects in ultra-fast digital designs, which are usually realized in some submicron technology, exhibit primarily analog character.

• **Analog design:** Behavior modeling capacity beyond SPICE is receiving more and more attention from analog designers, especially those involved in mixed-signal design, large analog design, and model development for new technologies. It is expected that an industry standard on analog hardware description languages will accelerate the art of analog design automation [2].

VHDL-A is an extension of VHDL, currently under development by the IEEE DASC 1076.1 Working Group, for the description and simulation of mixed analog and digital circuits and systems. The VHDL hardware description language was primarily designed for the modeling and the simulation of digital circuits [3][4]. During initial phases of the 1992 restandardization of VHDL, it was recognized, among other changes, that augmenting the language to handle analog designs was worthwhile [5]. Due to the overwhelming number of requirements already in consideration for enhanced digital functionality, the 1076 standardization committee elected to create a separate working group, officially numbered 1076.1, to deal with the problem of adding analog capabilities to VHDL. The intention was to decouple the analog requirements from the '92 language and to encapsulate them in a new standard that would be eventually merged with VHDL 1076 when the next restandardization occurs in 1997. As the time of this writing, it is more likely that VHDL-A will stay as a separate standard from VHDL 1076. It was also decided that the evolution of VHDL-A would follow the same process as for VHDL'92 [6] (e.g., requirements gathering, requirements analysis, language design, and balloting).

This article describes and analyzes the VHDL-A design requirements in terms of design objectives. Each design objective addresses one aspect of language design that is as concrete as possible. Each aspect is either a design goal that VHDL-A must meet or a design constraint that VHDL-A must obey. Each design objective is assigned a priority.

If there exists a conflict between different design objectives due to inherent technical constraints or if not all design objectives can be met within the given expected standardization time frame, then the language design emphasis will give to those design objectives of higher priority. Three orders of priority are identified as **must, should,** and **desirable** [7]:

• **must**　　　　The objective is an essential requirement and cannot be compromised. This means that the objective will be met in the first version of VHDL-A and in all subsequent revisions of the standard.

• **should**　　　The objective provides useful functionality beyond the basics and provides significant benefits if met. Unsatisfied **should** objectives can be used as a basis for requirements in subsequent re-standardizations.

• **desirable**　　The objective will enhance the usability and/or performance of the design. However, if not met, overall performance will not be severely compromised. Additionally, desirable objectives are to be used as a basis for new requirements in subsequent re-standardizations.

In formulating these design objectives, two guidelines are followed: First, design objectives related to the end user requirements must be stated as objectively as possible, independent of language design styles or approaches. Second, design objectives concerning the compatibility with VHDL must be specified carefully in order to give the VHDL-A language designers as much freedom as possible in pursuing high quality language design.

As a general guideline for the design of VHDL-A, design objectives must ensure a reasonable degree of portability, while giving enough freedom to the implementation. The idea is to not consciously close the door on something that we already know about, but that we don't plan to implement now just because it is not part of the current version of the norm. For example, one intention is to maintain portability without defining the simulation algorithm(s). However, this objective is much broader and covers more than analyses.

The emphasis of this article is to justify VHDL-A design objectives. To do so, design objectives are organized into four sections, concerning the scope of the language, the behavior description part, the structure description part, simulation-related part, and interface between analog and digital descriptions. Related design objectives are grouped, and appear in the section according to their relative importance in language design. The numbering of the design objectives is the same as in the official IEEE 1076.1 Design Objective Document [7], although they do not appear necessarily in the same order in this article. Interpretations or explanations to the design objectives are provided whenever necessary. In addition to the rationale for design objectives, certain implications to language design are also touched.

This article is structured as follows: Section 1.2 describes and justifies the scope of VHDL-A. Sections 1.3 and 1.4 describe the rationales for the structure and behavior description part of VHDL-A. Rationale for the simulation related aspect is presented in Section 1.5. Section 1.6 presents design objectives related to interface between analog and digital descriptions. Concluding remarks are made in Section 1.7.

1.2.   SCOPE OF VHDL-A

The nature of VHDL-A is a *description* and *simulation* language. The scope of VHDL-A defines which kind of systems it describes and which type of simulation it targets. In this section, we present design objectives related to the scope of VHDL-A. Also the relation with VHDL and the general guidelines in handling analog and digital descriptions are described here.

DO1   **(must)**   *VHDL-A must be suitable for the description and simulation of digital, analog, and mixed digital/analog systems at several abstraction levels (i.e. functional, behavioral, macromodel, and circuit levels for analog systems). VHDL-A must be able to support any design methodology and be technology independent.*

DO2   **(should)**   *VHDL-A should be suitable for the description and simulation of non-electrical (i.e. mechanical, thermal, etc.) and mixed electrical/non-electrical systems.*

DO3   **(must)**   *VHDL-A must be a "super-set" of VHDL'93. Any description that is valid in VHDL'93 must also be valid in VHDL-A, with the same simulation results. The only permitted exception to this is that new keywords introduced into the language may conflict with identifiers in VHDL'93.*

DO4   **(must)**   *VHDL-A must at least support time-domain analysis of lumped-element electrical systems.*

DO5   **(must)**   *Analog descriptions must re-use existing (VHDL'93) syntax where appropriate.*

DO6   **(should)**   *VHDL-A should support time-domain analysis of lumped-element non-electrical systems.*

DO7   **(should)**   *VHDL-A should support frequency-domain analysis of lumped-element electrical systems.*

DO8   **(should)**   *VHDL-A should support frequency-domain analysis of lumped-element non-electrical systems.*

DO9   **(should)**   *VHDL-A should not be restrictive in other types of analyses.*

DO10   **(must)**   *VHDL-A must provide mechanism(s) that allow the analog and digital behavioral parts of a mixed description to interact.*

DO11   **(must)**   *The basic A/D and D/A interaction mechanisms of VHDL-A must be completely defined and make no use of the "foreign interface" of VHDL'93.*

D012   **(should)**   *The model of the interface between the analog and digital descriptions should be customizable by the user.*

1.2.1.   Analog Hardware Description Language (DO1)

VHDL-A is intended to be an analog hardware description language (AHDL). As such, it must allow the direct specification of both the behavior and the structure of hardware systems. A primary motivation for analog hardware description languages is to support the modeling of physical systems. An AHDL must therefore allow to model the physical conservation laws, such as the energy conservation law which states that energy can neither be created nor destroyed, but it can only change its form. There are two fundamental mechanisms for modeling physical systems: behavioral modeling and structural modeling. Behavior modeling requires the language constructs to describe the physical laws of a system, in particular basic components of the systems, by mathematical equations or assignments. Structural modeling requires the language mechanisms to support hierarchial composition of basic components in a way to obey the physical laws. We will describe structural modeling and behavioral modeling respectively in Sections 1.3 and 1.4.

The distinction between analog and digital systems occurs whenever the designer selects how the system or parts of it will work. Both encompass different design methodologies and trade-offs. Roughly speaking, digital behavior is event-driven and therefore takes care of a limited number of well defined states, while analog behavior is continuous and takes care of entire waveforms.

To be consistent with the VHDL philosophy, VHDL-A must at least support the simulation of the specification. The standard VHDL simulation cycle uniquely defines the semantics of the language and is therefore an essential feature for portability. The aspects related to analog and mixed digital-analog simulation will be discussed later in this article.

It is very desirable that VHDL-A supports analog synthesis also. However, because analog synthesis is still in its infant stage, we do not know exactly whether the semantics for supporting synthesis differ significantly from the semantics for simulation. Therefore, it is not the objective of the current version of VHDL-A to support synthesis. However, whenever possible, VHDL-A must not make assumptions that may restrict its ability in supporting synthesis, or may restrict its future extensions for supporting synthesis. As a remark, particular emphasis must be given to support mixed-level abstractions, since it will be very helpful in supporting synthesis.

An AHDL standard is extremely useful for ASIC design industry. One of the major problems that designers face today is the consistency of their designs. No simulators on the market today can give them all the answers they need, so they have to use different tools, each requires different syntax for describing the same circuit. A standardized language is necessary to provide the design development, simulation, and data exchange of a hardware system.

1.2.2.   Abstraction Levels of Analog Circuits/Systems (DO1)

Depending on the complexity, the design will encompass several hierarchical descriptions at several levels of abstraction. Abstraction levels are today well defined for digital design [8]. A level of abstraction defines three aspects: the modeling concept, the timing model and the observable values. For instance, at the register transfer level, the modeling concept is guarded commands (reactive behavior), the timing model is discrete real time (clock ticks) and the observable values are interpreted bitstrings. It is usual to consider the abstraction levels from system (most abstract) to gate (less abstract) as pertaining to the digital world. For the most detailed level, the electrical, or circuit, level, the modeling concept is differential equations, the timing model is continuous real time and the observable values are continuous reals. This level is usually considered as to pertain to the analog world. It does not however cover all possible modeling methodologies as far as analog design is concerned.

Abstraction levels for analog design are less well established. In addition to the circuit level, other levels may be defined for more abstract analog descriptions. The main difference between digital and analog abstraction levels is that analog abstraction levels are based on continuous time, while digital ones are based on discrete time. Possible analog abstraction levels are, from the most abstract to the less abstract:

•**Functional level:** This modeling concept is based on signal-flow descriptions where a mathematical expression (e.g., a transfer function) maps the system input to the system output in a unidirectional way. Observable values are continuous values that usually do not obey conservation laws, such as Kirchhoff's laws for electrical systems.

•**Behavioral level:** This modeling concept is based on equations that describe the relations of constituent components. From this level down to circuit level, observable values are continuous and obey conservation laws such as Kirchhoff's voltage and current laws for electrical systems.

•**Macromodel level:** This modeling concept is based on the hierarchical assembly of lower level, circuit components, such as controlled sources. Only a very necessary number of these components are used to approximate the intended I/O behavior. The structure of a macromodel is usually simpler than the underlying hardware. Note that the macromodel level may not be considered as an abstraction level as such, but rather as a specific style of description.

•**Circuit level:** This modeling concept is based on the hierarchical assembly of primitive components such as resistors, capacitors, transistors, and sources, whose behaviors are pre-defined. The structure of a circuit description usually matches that of the underlying hardware.

•**Device level:** This modeling concept is based on two- or three-dimensional equations describing the behavior of one specific device (e.g. a transistor). Observable values are the electrical characteristics of the device, such as I/V characteristics.

•**Process level:** The modeling concept is based on quantic physics where aspects such as geometries and doping profiles are taken into account.

VHDL-A must support all levels of abstraction from the functional level to the circuit level (Figure 1). The device and process levels are beyond the scope of VHDL-A, because descriptions at these levels are too much detailed to be handled efficiently in the context of designing a whole system. Useful information from these levels are usually extracted into "equivalent," more abstract data expressed at the circuit level to perform analog or digital design. For example, detailed data from the device level is used to derive and validate circuit-level transistor models that can be handled by electrical simulators such as SPICE [9]. Another example is that the influence of the manufacturing process (at the process level) can be abstracted into statistical distributions of circuit component values (at the circuit level).

1.2.3. Design Methodology Independence (DO1)

A key to managing the complexity of a large design is to use top-down and bottom-up hierarchical design methodologies. Such designs usually contain mixed-level description. Further, the development of a large system may involve more than one team, and each team may base their design on different methodologies. Finally, the design of large systems may involve the use of existing design units. All these call for the need to support any design methodology, for which the ability to handle mixed levels of abstraction is essential.

A top-down approach starts from abstract descriptions that intend to capture essential aspects of the design while deliberately ignoring implementation details. These abstract descriptions are progressively refined until they reach a point where one implementation that satisfies the design requirements may be directly inferred from the description. A bottom-up approach takes the opposite way, by starting from a set of descriptions of pre-defined components and assembling them to form more complex components. In the EDA world, top-down design is usually associated with synthesis, while bottom-up design is usually associated with verification. A real design is usually done using mixed top-down and bottom-up approaches.

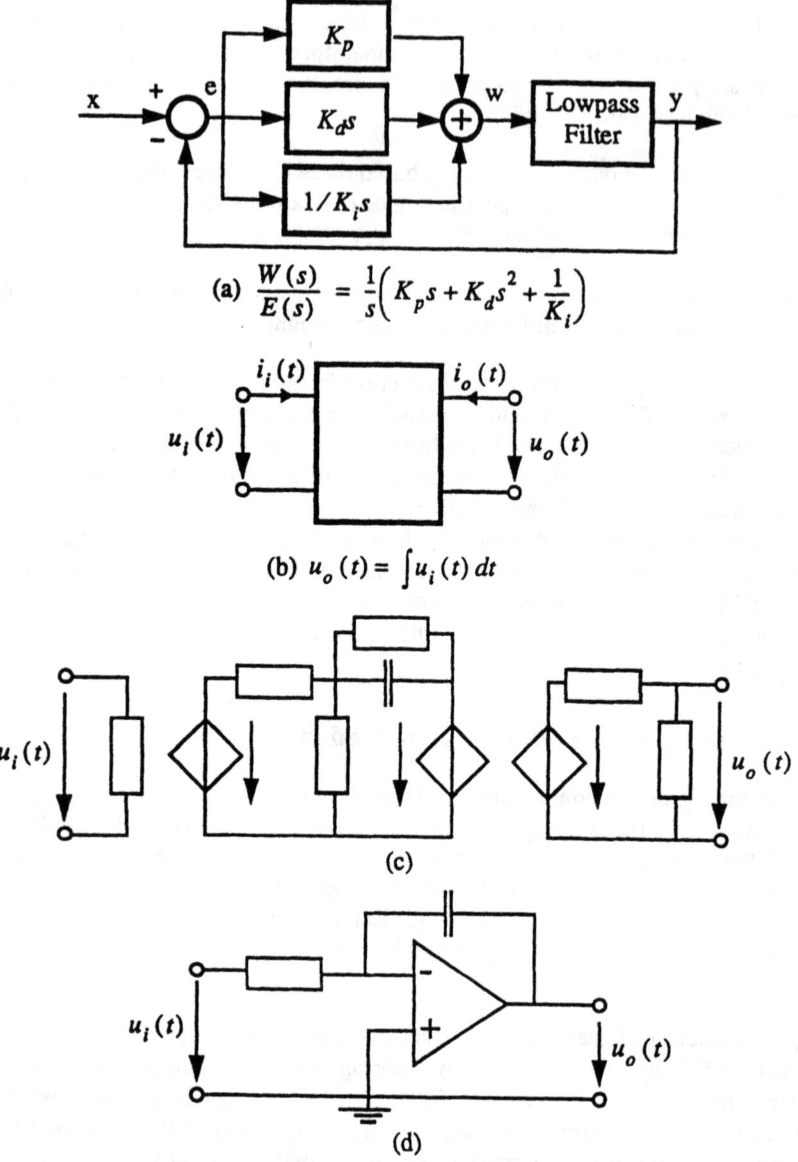

(a) $\dfrac{W(s)}{E(s)} = \dfrac{1}{s}\left(K_p s + K_d s^2 + \dfrac{1}{K_i} \right)$

(b) $u_o(t) = \int u_i(t)\,dt$

Figure 1: Abstraction levels supported by VHDL-A: (a) functional level, (b) behavioral level, (c) macromodel level, (d) circuit level (the opamp is described at the circuit level as well).

### 1.2.4.	Technology Independence (DO1)

There are three fundamental reasons for VHDL to be technology independent. First, as an international standard, VHDL-A is expected to support the description and simulation of the systems in different technologies. Second, with the advance of microelectronic technologies, a system (for example, a printed circuit board, a multiple-chip module, or even a chip) may contain different technologies. Third, and most important, technology independence is demanded from design re-use and technology migration. For example, an ASIC may be first designed using FPGA technology, then with PLD technology, and then mapped into gate arrays. It may even be redesigned later using full custom design technologies. The ability for VHDL-A to support the entire design spanning from the whole process is essential. This design objective further emphasizes the importance of supporting the description and simulation of mixed-level abstractions.

### 1.2.5.	Support of Mixed Analog and Digital Circuits/Systems (DO1, DO10, DO11, DO12)

As VHDL is primarily targeted towards the description and simulation of digital systems, VHDL-A must support the same objectives for analog and mixed analog-digital systems. Typical application areas of VHDL-A are conventional IC design, ASIC design, PCB applications, and electrical system design.

There are three basic reasons for hardware description languages to support the description and simulation of mixed analog and digital systems. First, a large portion of systems being designed nowadays, in particular ASICs, contain both digital and analog functionality. These are usually called mixed digital/analog circuits/systems. Second, high performance and high speed digital circuits exhibit analog characteristics. Their timing characteristics and power consumption usually cannot be calculated accurately at the digital abstraction levels. A full chip/system simulation requires that some part of the system be described at analog abstraction level, and some part be described at the digital abstraction level. Third, description of analog systems, particularly at the behavior level, may require techniques usually associated with the digital abstraction level. This is especially the case for certain non-electrical applications.

There are several implications to language design. First, direct interaction of analog and digital components is not possible because they are not handled by the same simulation kernel. By simulation kernel it is meant a standard mechanism that defines precisely how the VHDL-A statements have to be handled during simulation. Obviously, digital statements will not be handled the same way as analog statements, even if they are encapsulated in the same description. Therefore, VHDL-A assumes that two different simulation kernels exist, although this does not imply any particular implementation. Assuming two simulation kernels (one digital and one analog) is only necessary to formalize the semantics of analog objects and of the interactions between digital and analog objects that will be described in the VHDL-A language reference manual (1076.1 LRM). Given this, a mechanism must be established to translate signals from analog to digital and vice versa. Translation involves in general both a value conversion and a conversion between two timing models. These aspects will be discussed later in this article.

Second, care must be taken to ensure portability. VHDL-A will define a mixed simulation cycle based on A/D and D/A interactions. These interactions are not allowed to vary between implementations since they will be part of the norm. VHDL'93 provides a way to affect portability through the use of the "foreign interface" capability. VHDL-A must not make any use of such a feature, since it allows to defer the implementation of some part of the description outside VHDL-A. This is an open door to different implementations and therefore contradicts the portability objective.

Finally, VHDL-A should allow the user to decide where to use A/D and D/A conversion mechanisms and also how they will work (i.e. how the translation between signals will be performed) during simulation. This is necessary for VHDL-A to be technology independent.

1.2.6.   Support of Non-Electrical Systems (DO2)

Hardware systems that exhibit analog behavior are not confined to electrical circuits. Also, in many modern systems it is not sufficient to model and to validate only the behavior of the electrical part. In power applications, for instance, the thermal properties must be considered to get an accurate picture of how a system works. Other examples are electro-mechanical (mechatronic) systems, sensors, etc. It is also natural to extend the scope of VHDL-A to support non-electrical systems, since they are analogous to the electrical systems in their modeling requirements [10]. Differential equations that describe state variable solutions of systems in mechanical or thermal disciplines, are of the same form as differential equations that describe electrical systems.

Therefore, VHDL-A should support any system that can be described as a set of nonlinear ordinary differential and algebraic equations. This also includes thermal, mechanical, rotational, etc. systems in any combination (an example is shown in Figure2). Since there are commercial HDLs on the market that support this ability, customers expect that VHDL-A will do the same. Actually, the success of commercial HDLs provides the evidence that the ability of supporting mixed discipline systems is not only important, but also feasible.

There are two possible implications to language design. First, certain facilities may need to provide convenience to the user (for example, the ability to specify explicitly which discipline a component is). Second, care must be taken in language design so as not to use unnecessary assumptions that might limit the power of the language (for example, the use of specific electrical terminologies). Indeed, this is very much consistent with the VHDL design philosophy. A main feature of VHDL is to provide programming abilities, rather than concrete language constructs, for describing concrete components. For example, although VHDL is a digital hardware description language, however, no built-in constructs for logic gates are provided. This is part of the reason why VHDL has attracted attention from a wide range of application domains.

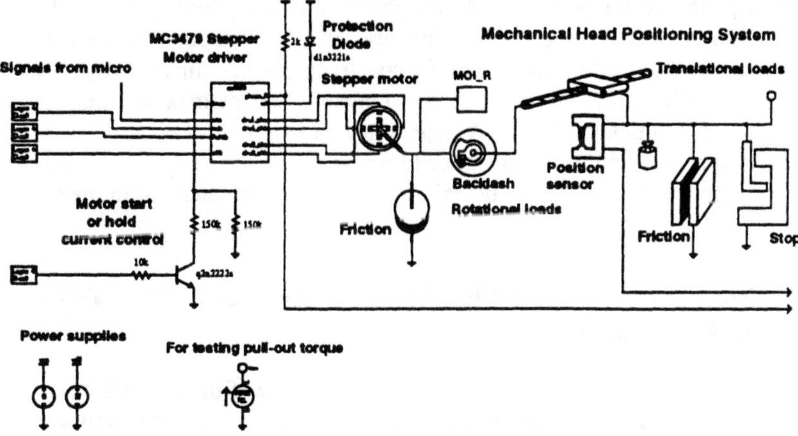

Figure 2: Example of a mechatronic system: Schematic diagram of the floppy disk drive head position controller [18]

1.2.7. Support and Re-Use of VHDL'93 (DO3, DO5)

VHDL is an IEEE standardized digital hardware description language. It has been widely accepted for the description and simulation of digital systems. Its applications are growing into various areas beyond its original design objectives (for example, synthesis and testing [14]). The development of VHDL itself also needs to consider analog modeling capacity in order to meet customer needs, especially for those customers in mixed system design.

VHDL has a rich ability in supporting large designs and design reuse, with techniques borrowed from software engineering. This ability is highly desirable to analog and mixed system designers. Thus, VHDL-A, built on top of VHDL, will gain additional advantages. Also, simulation is the most widely used and mature part of CAD. Simulation of systems with mixed abstractions, regardless whether they are mixed functionality systems, or mixed abstractions for the same functionality systems, calls for a unified language. Such a unified language is expected to give designers a friendly interface around which various simulation tools can be well integrated. A unified language is expected to be helpful in developing future simulation tools that may span a computer network in order to handle the continuously increasing need in simulation capacity.

VHDL-A will be designed as an extended VHDL (i.e., VHDL-A = VHDL'93 + analog and mixed digital-analog aspects). This is due to two considerations. First, to re-use those existing models and libraries developed in VHDL. Second, to ease the acceptance of VHDL-A by the digital design community. In the other words, we are not allowed to change existing VHDL unless we come up with a solution for reusing these existing design libraries and design tools. We also need to come up with a convincing and easy acceptable reason for existing VHDL users to re-learn their language. There are two basic interpretations of the superset concept here, which may affect fundamentally how VHDL-A language design proceeds. A straightforward interpretation is that VHDL-A keeps everything in VHDL (except for some keywords)—syntactic constructs and their semantic rule—with the addition of specific constructs for handling analog description and simulation. We can also interpret the superset concept in a broader sense: all constructs and semantics in VHDL-A degenerate to existing VHDL's if only digital description and simulation are concerned. In terms of the set concept, this means that not only the elements in the set can be enlarged, but also each element in the set can be enlarged.

1.2.8.   Support of Analog Simulation Techniques (DO4, DO6, DO7, DO8, DO9)

Analog simulation (usually called circuit simulation) is one of the oldest CAD activities in electronic design [11]. One of its main characteristics is that it encompasses many different kinds of analyses, as well as many different techniques to implement them in a simulator [12][13]. Time-domain circuit simulation is reportedly one of the most used, although the most costly, means to verify the behavior of an analog system without breadboarding. Moreover, mixed digital-analog modeling and simulation naturally requires having time-domain description and simulation capabilities from both digital and analog sides.

Assuming lumped-element descriptions of the system being simulated is also a consequence of the support of abstraction levels from circuit to behavioral level. From the behavior perspective, this assumption also implies that only ordinary differential equations (ODEs) are supported. As such, the microwave domain is not taken into account.

Given the wide use of the SPICE circuit simulator, it is natural for VHDL-A to support at least all the kinds of analysis SPICE supports, namely: DC, transient, small-signal AC, noise and distortion analyses. A time-domain description in VHDL-A, either behavioral or structural, should allow all these analyses, as SPICE already does from a netlist description.

Other types of analysis (especially power consumption, frequency-domain analysis, sensitivities, statistical, and harmonic balance) are desirable to be supported by VHDL-A. Care must be taken to ensure that these analyses can be supported as much as possible, because analog designers are not only interested in time-domain response, but other types of performances of equal importance in designing an analog system. However, the precondition here is that such considerations do not complicate existing VHDL syntax and semantics too much. So whenever possible, VHDL-A should use as few assumptions as possible that might restrict other types of analog analyses.

1.3. STRUCTURE ASPECTS

In this section, we describe and justify the design objectives that relate to the structure description in VHDL-A.

DO13   **(must)**   *VHDL-A must be capable of describing a digital, analog, and mixed digital/analog systems by structural composition of components.*

DO14   **(must)**   *Structural descriptions consisting of analog components of analog behaviors must observe conservation-law and/or signal-flow semantics.*

DO15   **(should)**   *VHDL-A should provide a migration path for SPICE based libraries and system descriptions (i.e. netlists).*

DO16   **(desirable)**   *It is desirable that VHDL-A provides a mechanism to specify netlists that are conditional upon the value of a constant parameter (conditional netlists). A special case of this is collapsing two nodes.*

DO17   **(desirable)**   *It is desirable that VHDL-A provides a mechanism to specify regular array structures of instances with constant dimensions.*

1.3.1. Structure Description: Conservation-Law and Signal-Flow Semantics (DO13, DO14)

A system is usually built from a connected network of *components*. Depending on the complexity of the system, components may be further decomposed hierarchically into another set of interconnected, lower-level components. Finally, leaf cell components have to encapsulate some behavior if the description is intended to be simulated. Connection points in the network of an analog system are called *nodes*. Terminals of components are called *pins*. There are two types of networks—*conservation-law* and *signal-flow*, which are dependent on the *semantics of connection*.

An example of a conservation-law network is an electronic circuit. In a conservation-law network, there is a so-called *through quantity* (current for electronic circuits) associated with each *branch*, and a so-called *across quantity* (voltage for electronic circuits) associated with each node. Usually, but not necessarily, the product of the through quantity by the across quantity is a power. This will depend on the modeling requirements [10]. For example, a model of a mechanical system may be written using through quantity to be the force and across quantity to be the position. The product across-by-through is work in that case. Conservation-law networks also have specific connection semantics, such as Kirchhoff's laws for electronic circuits. For each node in such a network, the algebraic sum of through quantities associated with all the branches entering that node must be zero (KCL for electronic circuits). For each *loop*, the sum of the across quantity drops along the loop is zero (KVL for electronic circuits). Conservation-law networks are also directionless: connecting two components together implies a mutually-shared contribution to each other's behavior. Conservation-law networks are supported from the circuit abstraction level up to the behavioral abstraction level. Moreover, they are required to support the description of lumped-element systems (either electrical or from other disciplines).

Examples of signal-flow networks can be found in the early stage of an electronic system design or in a control system design. In a signal-flow network, quantities are associated with nodes only. Signal-flow semantics implies that the terminals to which different components connect have the same quantity. It also implies a directed interaction between nodes. Signal-flow networks are useful to describe linear systems. A more general form is block diagram description, which may represent any kind of nonlinear operation to be performed on quantities entering a block [10]. Signal-flow networks are supported at the functional abstraction level.

In addition to the high-level description of analog systems, signal-flow semantics is required to model the phenomena that one component is controlled by a quantity associated with terminals of another component. Take a voltage-controlled current source $I=f(V)$ for an example. Suppose that V is the voltage of a terminal of COMPONENT_A and the voltage-controlled current source is used inside COMPONENT_B. Then we have to import V into COMPONENT_A through its interface. We could have something like COMPONENT_B(V, pin1, pin2,...), where pin1 and pin2 are two electrical terminals of COMPONENT_B. Here, binding for V obeys the signal-flow semantics, where bindings for pin1 and pin2 obey the conservation-law semantics.

Signal-flow semantics provides the simplest binding rule for analog system specification. However, it is not supported as a generally available structure mechanism in SPICE-like simulators, which are originally developed with the emphasis on integrated circuits consisting of a limited number of primitive components. Incorporation of the signal-flow semantics into VHDL-A will not complicate the language, but give rise to more power, and lead to more applications, such as control system simulation and possibly analog synthesis.

1.3.2.   SPICE Compatibility (DO15)

SPICE is undoubtedly the most widely used analog simulator. Its associated syntax has hence became a *de facto* standard, although many flavors of it exist on the market. The acceptance of VHDL-A will then depend on how the language will allow the SPICE users to re-use existing SPICE netlists. However, the objective is neither to include a SPICE netlist format (actually which one?) in VHDL-A, nor to allow VHDL-A to *understand* it. VHDL-A should be powerful enough to allow the translation of a SPICE netlist into a VHDL-A description, possibly through the use of automatic tools (it is always easier to translate a simple format into a more complex one than the other way around).

However, nothing prevents future VHDL-A tools from using SPICE netlists directly. Some VHDL tools provide this feature to allow mixed mode modeling and simulation without any modification on the language itself [15]. The use of VHDL'93 foreign interface mechanism or the development of specific dedicated packages should also help.

Another issue is how to handle device models that are hard coded in SPICE-like simulators. Many device models are built into simulators, because no standard procedural interface exists in SPICE-like simulators. Programming constructs of VHDL-A should allow these models to be rewritten in VHDL-A (see design objectives in Section 1.4). On the other hand, it is desirable that VHDL-A allows to incorporate device models written in programming languages. This may be achieved by using VHDL'93 foreign interface.

1.3.3.   Conditional Netlists (DO16)

The need to switch netlist description for a component arises in the analog design process. Adding drain and source parasitic resistances into an ideal MOS model are a typical example: a non-zero value forces the netlist to have one or two supplementary nodes. Node collapsing is the opposite situation (due to a short-circuit for example). Higher level models should also gain flexibility from conditional netlisting.

Switching between a behavioral model and a netlist model is also desirable, since behavior and structure could be freely mixed in the same description, which VHDL already allows.

An issue here is whether VHDL-A supports this functionality between different simulation runs (before simulation, or in VHDL terminology at the elaboration time) or during a simulation run. It is believed that support of an arbitrary conditional netlist during simulation is far away from the original VHDL design objective. It has therefore been decided not to include it as a VHDL-A objective.

1.3.4.   Regular Structures (DO17)

Flexibility (or generality) is a guideline VHDL has already adopted. It is hence natural to extend it to analog and mixed digital-analog structures. As an example, the ability to specify regular analog array structure is useful for designing such circuits as artificial neural networks.

1.4.   BEHAVIOR ASPECTS

VHDL-A with only the structural description mechanism described in Section 1.3 will basically provide what SPICE provides, although in the more flexible context of VHDL. The development of models for complex systems would still require the use of predefined low level components in a bottom-up manner, which inherently limits the descriptive capabilities of the language. Therefore, a major goal of VHDL-A is to support behavioral modeling (i.e., describing models to a simulator by describing mathematical equations directly). In this section, we describe and justify the design objectives that relate to the behavior description in VHDL-A.

DO18   **(must)**   *VHDL-A must support behavior specification of an analog system by a set of linear/nonlinear differential/algebraic equations and/or by a sequence of assignments.*

DO19   **(must)**   *VHDL-A must support the description of continuous time behavior both by explicit and by implicit equations.*

DO20   **(must)**   *VHDL-A must support mixed structural and behavior specification for a single analog component. For this purpose, VHDL-A must provide a mechanism for the behavior specification portion to access quantities associated with the structural specification portion and vice versa.*

DO21   **(must)**   *VHDL-A must provide a mechanism to access quantities that are used inside an analog component but not associated with terminals (i.e. connection points satisfying the KCL/KVL semantics) from outside of the component.*

DO22   **(must)**   *VHDL-A must support a mechanism to parameterize a model. The support must at least include parameters that are constants. It is desirable to support parameters that vary during simulation. It must be possible to distinguish parameters that intentionally have been left unspecified from parameters having their default initial value.*

DO23   **(must)**   *VHDL-A must support the description of analog behavior that depends on the independent variable of an analysis.*

DO24   **(must)**   *VHDL-A must support discontinuous analog behavior and waveforms. In particular, it must be possible to enforce an analog time step in response to some change in the analog part.*

DO25   **(must)**   *The maximum simulation time for a mixed analog/digital system must not be restricted by the minimum time resolution of VHDL'93.*

DO26   **(must)**   *VHDL-A must provide a time derivative operator.*

DO27   **(must)**   *VHDL-A must provide pre-defined mathematical functions (such as sine, cosine, exp, log, ln, etc.). It should also provide a way to define custom mathematical functions.*

DO28   **(should)**   *VHDL-A should support the description of piecewise defined behavior.*

DO29   **(should)**   *VHDL-A should provide an integral operator. It must be at least defined from time zero to the current simulation time.*

DO30   **(should)**   *VHDL-A should support the annotation of physical units to waveform.*

DO31   **(desirable)**   *It is desirable that VHDL-A support dimensional analysis.*

1.4.1.   Behavior Description (DO18, DO19)

A primary motivation for an analog hardware description language such as VHDL-A is to make it easier for the user to create models for new devices (for example, in VLSI circuits using the deep submicron technology). Another goal is to help develop simple behavior models for portions of complex circuits or the environment of ASIC applications (for example, mechanical parts [16] or thermal environment [17]).

Behavior modeling is a more general concept than commonly used ideal behavior modeling and transfer-function modeling [20]. SPICE-like simulators *do not support* behavior modeling. With SPICE-like simulators, the model must represent the behavior by a netlist of built-in primitive components, mostly controlled sources with possibly polynomial behavior. This is often called the *macromodeling* approach, which is intended to get a compact representation of a circuit that captures features that are relevant to a particular purpose, while deliberately ignoring unnecessary information [21][22]. Although macromodeling has been widely used before the advent of AHDLs, it is not always an easy task, and in some cases, it cannot be used at all. Another point is that a macromodel depends heavily on the underlying component library, which may cause portability problems. However, VHDL-A will support macromodeling as a direct consequence of its support of structure and behavior descriptions.

True behavior modeling [19] as supported in VHDL-A should allow the user to describe a model according to design decisions. An example of behavior specification is by signal-flow diagrams or design equations [2]. In some cases, describing the behavior as a set of linear/nonlinear differential/algebraic equations is more straightforward (and more compact) than using an equivalent macromodel.

Behavior modeling is also related to simulation issues. Simulating a behavior specification of a circuit/system to design (or a circuit/system with portions available as behavior specification) may be less expensive in term of simulation resources. It may even become an absolute necessity to perform full system simulation.

1.4.2.   Support of Equations (DO18, DO19)

A natural way to describe the behavior of an analog system is to use mathematical equations. An equation describes a relation (or called a constraint) on a set of quantities that an analog system must always satisfy. An equation can be said to be in either implicit or explicit form. An *implicit* equation may take the *root form* as

$$(1) \qquad F(Q_1,Q_2,...,P_1,P_2,...,V_1,V_2,...) = 0$$

or take the *fixed-point form* as

$$(2) \qquad Q_i = F(Q_1,Q_2,...,Q_i,P_1,P_2,...,V_1,V_2,...)$$

where F denotes an expression involving quantities or time derivative of quantities of any order, Q_i are quantities involved in the equation (unknowns), P_i are constant parameters, and V_i are the independent variables (e.g. time or frequency). As an example, here is the implicit equation binding two currents in a current mirror circuit:

$$(3) \qquad U_t \cdot \log\left(\frac{I}{I_{ref}}\right) = -RI$$

The solution of implicit equations can be computed through iterations, providing that initial guesses are given for the unknown quantities. Note that stating this does not imply any particular iterative method to actually perform the computation.

An *explicit*, or *separable*, or *closed-form*, equation have the following general form:

$$(4) \qquad Q_1 = G(Q_2,Q_3,...,P_1,P_2,...,V_1,V_2,...)$$

where quantity Q_1 does not appear on the right-hand side of the equation. A very simple example of explicit equation is the ideal constitutive equation of a capacitor:

$$(5) \qquad i(t) = C\frac{d}{dt}u(t)$$

Given the values of the quantities appeared in the right-hand side of an explicit equation, the quantity on the left-hand side can be computed immediately.

Both implicit and explicit forms must be supported in VHDL-A, because both may naturally occur when describing analog behavior. While it is straightforward to transform an explicit equation into an implicit one through algebraic manipulations, the other way around, i.e. implicit to explicit, is not always possible, or may lead to less understandable descriptions.

Conceptually, all equations are simultaneous and must be gathered into one set before simulation. The resulting set of equations, which will be actually solved by the simulator, depends on the formulation method used by the simulator and on the underlying connection semantics. The order in which equations are written in the description has no effect on the simulation results. Writing behavioral models using mathematical equations is called the *equational style* of behavior description.

1.4.3.   Support of Assignments (DO18)

Certain analog behavior may be described as a sequence of assignments; this is called
the *procedural style*. For example, an ideal differential amplifier can be described is given
in Figure 3 (the syntax used is only illustrative). In the case where complex
computations are needed (e.g., in a conditional branch), they have to be handled in a
sequential way.

```
if deltaV > deltaVMax then
    iOut := iMax
elsif deltaV < -deltaVMax then
    iOut := -iMax
else
    iOut := Gm * deltaV
end if;
```

Figure 3:   Analog behavior as a sequence of assignments.

Procedural style might be required when some computation of equation parameters is
needed. Also, explicit equations may be solved procedurally to accelerate the
simulation.

1.4.4.   Support of Mixed Behavioral and Structural Descriptions (DO20, DO21)

The objective to support mixed behavioral and structural descriptions comes from a
fundamental requirement that VHDL-A must support evolutionary design
methodologies spanning the whole design process. For such design methodologies, it is
often the case that a single design can contain aspects that are structural in nature and
others that are behavioral in nature. This type of description tends to evolve naturally
through the process of hierarchical design of a part: as the layers of the part are
successively designed, more and more of the description may be given as a netlist of
component instantiation; the remaining parts are still described as behavioral
specification. On the other hand, during the bottom-up verification phase, it is a "must"
that the behavior of leaf cells be described at the behavior level in order to simulate the
entire system. Also, in order to do full-system verification, it is often required to model
some part of the design directly at the behavior level. This ability is also desirable for
simulation of mixed discipline systems where some part may have to be described at the
behavioral level.

There are several reasons for supporting mixed structure and behavior description at a
single design entity. First, VHDL'93 allows the mixture of behavior and structure
description for one single design entity, and so are some commercial HDLs. Second, the
end users may want to perform some checking or statistics collection regarding a design
described at the structural level. For example, interface timing checking for digital
circuits and power consumption for analog circuits. These checking tasks require the
access of the behaviors of the systems described at the structural level. Third, under
some circumstances, it may not be a good practice to represent a single design entity

that has some part behavioral and some part structural in two separate entities. Finally, it is believed that this ability should provide more convenience.

Since both behavior and structure may be freely mixed in the same description, the behavior part may need to have access (read, write, or both read and write) to quantities that are used inside an analog component. The interface of an analog component usually defines connection points, or pins (i.e. terminals at which conservation laws are satisfied). However, abstract models may describe some coupling between behavior and structure that does not rely on conservation-law connection points. Controlled sources are a typical example: a current source in a design entity A with its current value controlled by the current of a voltage source in another design entity B. The current of the voltage source is a through quantity not associated with terminals of entity B. A convenient and natural way must be provided the user to describe the voltage source, entity B, and entity A. In particular, it is not desirable to specify this voltage source as a three-terminal device. Note that the values involved in such coupling may change during the same simulation run.

1.4.5.   Support of Parametric Models (DO22, DO23)

Parameters usually refer to those inputs to a model (circuit primitive) that are not related to its connection (for example, resistance value of a resistor, temperature for a transistor model). Such values are known before simulation and they are not allowed to change during one simulation run.

However, it may be possible that one model's parameters are through/across quantities of another model from a different discipline. For example, temperature of a transistor may be an across quantity in the thermal network used to model the thermal environment of the transistor (Figure 4) [17]. The values of such *parameters* will change during simulation.

A special case is parameters that depend on the independent variable of an analysis. For example, time-dependent resistance in transient analysis, and frequency-dependent capacitance in frequency analysis. This implies that the independent variable should be available in a model.

There are several requirements for supporting parametric models: First, parameters can be set to default values. Second, parameters can be deliberately left unspecified. In that case, they act as flags to indicate that some additional work is to be performed. A typical example is the preprocessing of MOS model parameters. Third, parameters may be analysis specific or generally applicable. Fourth, parameters may be not necessarily coefficients for equations or condition variables. They may be those used to create coefficients for equations or condition variables. The ability of value or range checking for parameter must be supported.

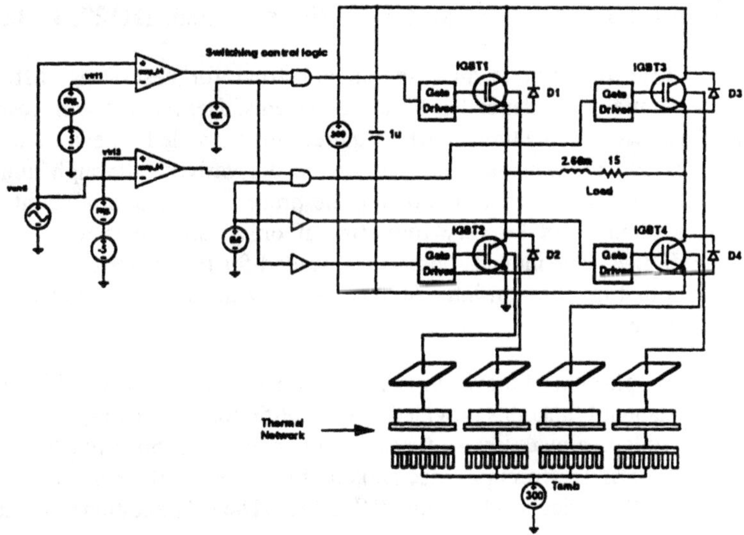

Figure 4:   Example of an electro-thermal network: The basic electro-thermal network used for the PWM inverter simulations [18]

1.4.6.   Support Piecewise Defined Behavior (DO24, DO28)

Analog behavioral modeling may require to define particular regions of operations exhibiting different behaviors (described by different set of equations). Examples are modeling the behavior of a MOS transistor in the electrical domain or modeling the behavior of a bouncing ball, or of stick/slip friction, in the mechanical domain. The conditional switching of equations during simulation must hence be supported in VHDL-A. Care must be taken, however, to ensure correct simulation results at the boundary of each region of operation. This may be done either explicitly in the model or implicitly in the simulator.

Another kind of piecewise defined behavior is the explicit specification of piecewise linear input stimuli. It is common in SPICE-like simulators to take the breakpoints into account during simulation by forcing an evaluation of the state of the circuit when the break occurs in time. This does not, however, affect how the model is written, but rather how accurately the simulator will compute the output waveforms. The issue here is to decide whether or not the new canonical mixed-mode simulation cycle in VHDL-A will explicitly define what to do in this case.

Yet another aspect that requires the support of discontinuous behavior is mixed digital-analog description and simulation. It will be discussed later in this article.

1.4.7. Standard Operators and Functions (DO26, DO27, DO29)

The time derivative operator is an essential construct for describing differential equations of dynamic (time-dependent) analog systems. Derivatives with respect to other variable/quantity than the time may be required (for example for device modeling). Such derivatives can be expressed by using time derivatives, although numerical inaccuracy may occur in that case. Following the preceding discussion about SPICE compatibility, it should not be very limitating if only derivatives w.r.t. time are supported in VHDL-A. The integral operator is needed for the matter of convenience. Every expression that involves an integral may be easily transformed into another one that uses a derivative.

To facilitate the description of a system, certain mathematical tools and abilities must be provided. All standard functions provided in a scientific (or engineering) language like C should be available in a math library for the user convenience. Such functions should not require the VHDL'93 "foreign" mechanism, since they will be provided by the coming standard mathematical package for VHDL [25]. The only additional requirement for VHDL-A is to allow the future standard mathematical functions to operate on double precision floating values. Other functions like NAG or IMSL are outside of the scope of VHDL-A, but anybody who needs these functions can write a package to define them using VHDL'93 "foreign" mechanism. It may also be useful for a user to define their own functions, or to overload existing ones with proprietary implementations.

1.4.8. Support of Physical Units and Dimensions (DO30, DO31)

All physical quantities are associated with units of measurement. There are seven basic units: meter, kilogram, second, ampere, kelvin, candela, and mole, as defined by the standard ISO-1000 (hence called SI units). All other units can be expressed in terms of these seven basic units.

VHDL-A must support the ability of manipulating a physical quantity together with its related unit. This is very useful for documentation purposes and for interpreting the simulation results, in particular for mixed discipline systems.

Further, it is very desirable that any assignment operation or equation formulation be checked for dimensional equivalence. Equivalence checking is a more involved process, which requires defining and applying rules that specify how dimensioned quantities are arithmetically combined and how basic units are combined in terms of non-basic units (for example, use the watt as the unit of power).

Dimensional analysis is available in MHDL [23] and Ada [24]. It is considered as desirable design objective in VHDL-A, since it is not needed as a basic functionality. On the other hand, the annotation of quantities with physical units should be straightforward to include in VHDL-A, and thus has been considered as a design objective that must be met in the first version of VHDL-A. Care must be taken, however, not to affect the future implementation of dimensional analysis.

1.5.   SIMULATION MECHANISMS

This section is devoted to design objectives that are targeted for simulation specific issues. There are three fundamental reasons for these design objectives. First, VHDL, which is a basis of VHDL-A, defines a canonical simulation cycle. In order to be compatible with VHDL, VHDL-A must define its simulation mechanism that degenerates to the canonical simulation cycle of VHDL for pure digital systems. Second, the scope of VHDL-A is to support the simulation of mixed digital and analog systems. Therefore, the interaction between digital simulation and the analog simulation, has to be specified. Finally, analog simulation (and analysis) itself requires to specify certain parameters (for example, the convergence tolerance for the Newton-Raphson algorithm). However, these parameters depend not only on specific analog analysis, but also on the specific algorithm, and even on specific implementation of the algorithm in a specific tool.

DO37   **(must)**   *Compliance with VHDL-A must not depend on the specification of an underlying simulation algorithm for the analog kernel.*

DO38   **(must)**   *VHDL-A must define the mechanism related to the simulation of the analog part of a VHDL-A description, as well as the mechanism related to the simulation of mixed digital-analog descriptions.*

DO39   **(must)**   *VHDL-A must support the specification of user-defined initial conditions. It must also provide a way to start the simulation in a consistent state.*

DO25   **(must)**   *The maximum simulation time for a mixed analog/digital system must not be restricted by the minimum time resolution of VHDL'93.*

DO40   **(should)**   *A communication mechanism should be provided to pass information back and forth between the VHDL-A model and the analog simulation kernel. Only the communication mechanism should be defined, not the way the simulator handles them.*

1.5.1.   Support of Analog Simulation and Mixed Analog-Digital Simulation (DO37, DO38)

For a digital system, the specification of its behavior and structure in VHDL also defines uniquely how the system is simulated (i.e., the simulation algorithm). Analog description is only denotational, which specifies only the set of equations that the system must satisfy. There are many algorithms for solving the set of equations and finding the behavior of the system. Each has its advantage for simulating certain circuits/systems under certain design styles and targeted technologies. Depending on the needs, there may be a choice between a very accurate, but expensive (time and memory consuming), and one that is simplified, but fast. VHDL-A, targeted as a language standard, must be technology-independent and methodology-independent. Therefore, no specific algorithm can be defined or included in the new mixed digital-analog canonical simulation cycle. In fact, this design objective leads to one of the most challenging aspects of VHDL-A design.

On the other hand, we need to support the simulation of mixed analog and digital systems. Therefore, VHDL-A must provide a mechanism related to mixed analog and digital simulation. Ideally, we would like to have an abstract model that has two essential features: one is to serve as a common abstraction of all analog simulators, and the other is for the integration with digital simulation (i.e., time synchronization [26]). These aspects are very important to ensure a reasonable degree of portability for the models written in VHDL-A.

Mechanisms to support mixed analog and digital simulation and in particular the interaction between analog and digital simulation kernels, is perhaps the most critical but least known aspect of VHDL-A. From the analog description and simulation point of view, this is a language design, and even a simulator-development issue. However, digital behavior specification also defines how it simulates. For the sake of compatibility and to support mixed simulation, the interaction between the two simulation kernels is a design objective issue.

1.5.2.   Initialization of Analog Simulation (DO39)

Initialization is a part of the digital canonical simulation cycle in VHDL'93. Initialization of digital part of a mixed digital and analog system requires the initial states of analog part to be known. Therefore for both the compatibility and functionality, VHDL-A should define explicitly the initialization of analog part. However, analog initialization is a much more complicated process than digital initialization.

In VHDL'93, initialization is the first step in the canonical simulation cycle[2]. It mainly assigns initial values to all signals and variables in the description, and runs each process once until it suspends. A digital model is usually in an *inconsistent* state after initialization. Consistency would only be reached, if the processes were executed repeatedly until the event queue is empty, while at the same time the stimuli are frozen (i.e., no new events are created).

Analog initialization, on the other hand, involves the formulation of system equations and the solution of the system of equations in order to find an initial state (DC

---

[2] To be exact, the VHDL'93 LRM states: "The execution of a model consists of an initialization phase followed by the repetitive execution of process statements in the description of that model. Each such repetition is said to be a *simulation cycle*".

operating point computation). It is commonly accepted that the formulation of system equations and its solution is beyond the scope of VHDL-A. So what we can probably define as a design objective is that VHDL-A must support the specification of initial conditions. A DC analysis is another, specific kind of analysis that may be required to reach a consistent state before time-domain or small-signal AC analysis can be performed.

One concern in the specification of initial conditions is how to handle inconsistent initial conditions, especially for mixed analog and digital systems. In addition, there are two other concepts related to initial conditions, which may need to be considered altogether during language design. The first one is starting point of iterative algorithms for solving nonlinear equations. For some analog simulators, initial conditions are also used as the starting point for numerical solution algorithms. The other concept is that, for certain analog systems that have multiple DC states (solutions), some information about the initial state must be provided in order to find a unique solution.

1.5.3.   Simulation Time Limitation (DO25)

This objective addresses a problem that may occur during the simulation of mixed digital-analog models. If analog and digital kernels are synchronized with the VHDL'93 minimum resolvable time (MRT) of 1fs, the maximum simulation time of a mixed system is restricted to about 9s if the time is implemented as a double precision variable (53 bits of resolution => 2**53 * 1fs maximum time). This is clearly insufficient for some system applications where the time constants may easily be in the order of seconds (e.g. if a motor is involved). VHDL provides a workaround to this problem by allowing to select a secondary unit of time, which is larger than 1fs, as the MRT at the price of a reduction of accuracy. VHDL-A must also support this feature at the A/D interfaces of the model.

For pure analog models, simulation time limitation should not be a problem provided that time is based on real values, instead of on integer values as used in digital simulation. Moreover, the problem of stiff systems (i.e., analog systems exhibiting time constants of different order of magnitude) is usually solved in the simulator by using appropriate numerical integration methods (e.g., Gear methods). This does not have any impact on the VHDL-A language as such.

1.5.4.   Simulation Control (DO40)

Simulation control is needed in some form, because analog simulation methods may use various numerical algorithms. The accuracy of the solution is usually related to a set of numerical tolerances, which are used to decide wether the solution has converged or not, and also to the actual implementation of the numerical algorithms (e.g. various limiting schemes to avoid numerical overflow). Another point related to simulation control is to allow to perform statistical simulations, where the same model is simulated several times for different values of some specific parameters.

Simulation control is a controversial aspect since it may be considered as a tool issue that should be solved outside of the language. It is indeed an objective of VHDL-A to be independent of any implementation (i.e. of any specific simulation algorithm). DO40, however, simply states that some communication mechanism should exist in VHDL-A to let the model provide information to the simulator, whatever it is, and conversely to allow the model to get simulation related information (even if this might be considered to be bad modeling practice). It will eventually be up to the language design task to figure out if such a feature is really needed in VHDL-A.

1.6. INTERFACE BETWEEN ANALOG AND DIGITAL DESCRIPTIONS

Simulation of mixed analog and digital systems requires three aspects of analog/digital interactions to be specified as design objectives. One is the dynamic communication between the two different simulation kernels that handle respectively analog and digital abstractions. This has been described in the previous section. The other two aspects are rather static, concerning the conversion between digital signal and analog waveform (behavior), and the conversion between digital connection point and analog connection point (structure). In fact, the interaction between simulation kernels is embodied through the conversion between analog and digital abstractions.

DO32 **(should)** *Default conversions between analog and digital connection points should be provided.*

DO33 **(must)** *A system-supplied thresholding function must be accessible to the user in defining customized thresholding functions. The user must be able to define what a "significant change" is that defines thresholds within the analog system.*

DO34 **(must)** *VHDL-A must support at least the following analog to digital interaction mechanisms:*
- a digital process must be able to read an analog value
- it must be possible to make a digital process sensitive to an analog value

DO35 **(must)** *VHDL-A must support at least the following digital to analog interaction mechanisms:*
- an analog behavioral model must be able to read a digital signal
- it must be possible to make an analog behavioral model sensitive to a digital signal. In particular, it must be possible to enforce an analog time step in response to an event on a digital signal.

DO36　**(should)**　　　*VHDL-A should addresses the problem of discontinuities the digital part may produce due to the intrinsic properties of digital signals.*

1.6.1.　Default Conversions between Analog and Digital Structures (DO32)

Due to the strong typing mechanism of VHDL, it would not be allowed to directly connect a VHDL signal to a VHDL-A pin, or a VHDL-A node to a VHDL port. Also, it is not in the VHDL philosophy to define default conversion functions to allow to connect two ports of different types. Moreover, type casting is limited to a few cases. However, there are cases in where an analog-digital interface is needed only to simulate parts of a system at different abstraction levels. For example, when the real system will work as a digital system, but some detailed, critical timing information is needed to validate global behavior. In that case, there won't be any real A/D or D/A converter in the final system, but simulation needs some conversion to proceed. The objective here, although not a primary one, is just to avoid forcing the user to describe explicit conversion statements whenever they do not correspond to real functionality in the designed system.

1.6.2.　Analog to Digital Interactions (DO33, DO34)

Analog to digital interaction involves the transformation of an analog waveform, which is continuous in amplitude and in time, into a digital signal, which is quantized in amplitude and discrete in time. Therefore both a value conversion and a time conversion are to be performed.

For data value conversion, a threshold function is usually used to convert an analog waveform to a digital signal. For an electronic system, the voltage of the analog connection point would be transformed into a legal digital value in the logic value system in use. The current of the analog connection point may also be used to get a more specific digital value when multi-valued logic is used (such as the IEEE 1164 standard 9-state LVS). From the analog side, the analog part should see a load (e.g. a stray capacitance) that models the fact that it is connected to something and should not be left open. For time value conversion, the real based analog time has to be rounded or truncated to the nearest multiple of the digital integer based time.

What the threshold is must be customizable by the user. The mechanism to specify that the digital part is waiting for some event from the analog part, along with how the simulation should react to that event, must be precisely defined in VHDL-A.

1.6.3.　Digital to Analog Interaction (DO35, DO36)

Digital to analog interaction involves the transformation of a digital signal into an analog waveform. It also requires a value conversion and a time conversion.

For data value conversion, the value of the digital signal has to be translated into *both* across and through values in the analog part. For electronic systems, a specific voltage-controlled switch might be used [26]. Depending on the signal value, the switch selects one of several identical subcircuits (usually a non-ideal voltage source with a parallel capacitor), but with different component values, to act as an input component for the analog part. The number of input subcircuits depends on the number of states the digital signal may take. From the digital part, the fact that it is not open but connected to something might be modelled as a specific assignment delay for the signal going out the digital port. For the time value conversion, the integer based digital value is merely converted into its equivalent floating representation.

An important aspect is the possible discontinuities of an analog waveform converted from a digital signal. The instantaneous value change of a digital signal is likely to produce numerical difficulties when computing the state of the analog part in reaction to this change. Some "smoothing" function should be defined to avoid this. At minimum, a piecewise linear "analog delay" should be produced on the analog waveforms at the D/A interface. More complex behaviors are also possible, such as an exponential change between two analog values.

1.7.   CONCLUDING REMARKS

In this paper, the complete set of design objectives for the design of the VHDL-A language have been given and discussed. All the related work has been done, and is currently going on, within the IEEE 1076.1 Working Group. The goal of this group is to propose a consistent set of extensions that support the description and simulation of analog and mixed analog-digital hardware systems. Based on the design objectives, a global architecture for VHDL-A is soon to be completed. It will contain precise definitions of all the new concepts VHDL-A is bringing to VHDL and will ensure that the VHDL philosophy is not violated. This last point is undoubtedly a strong constraint that limits the space of possible implementations. MHDL, another approach to provide an analog hardware description language but more focused on the microwave domain, did not have such a restriction to design the language (see also this issue). Nevertheless, the constraint might be an advantage, since VHDL-A is evolving from VHDL—a standard and successful HDL.

1.7.1.   Acknowledgments

This article is based on a report prepared for United States Air Force Rome Laboratory and on the Design Objective Document, a working document of the IEEE 1076.1 Working Group. The authors would like to thank Rome Laboratory for allowing writing this paper for public use. The elaboration of the VHDL-A design objectives was possible thanks to the contributions of many people in the 1076.1 Working Group, especially Jean-Michel Bergé of CNET, Ernst Christen of Analogy, and Dan FitzPatrick of Cadence. The authors would like to thank Mike Fiegenbaum of Analogy and anonymous reviewers for their comments that improve the presentation of this paper.

REFERENCES

[1]   A. Yang and K. Mayaram, "Simulation and modeling: Stepping beyond SPICE", IEEE Circuits and Devices, vol. 10, no. 3, p. 13, May 1994.

[2]   R.A. Rutenbar, "Analog design automation: where are we? where are we going?", in Proc. IEEE Custom Integrated Circuits Conference, 1993, pp.13.1.1-13.1.8.

[3]   VHDL special issue of IEEE Design & Test of Computers magazine, April 1986.

[4]   IEEE Standard VHDL Language Reference Manual, IEEE Standard 1076-1993, SH16840, IEEE Press: Piscataway, NJ, 1993.

[5]   S. Krolikoski and J. Mermet, "The future of the IEEE VHDL Analysis and Standardization Group", Proc. Euro-VHDL'91, Stockholm, September 1991, pp. 104-110.

[6]   M. Shahdad, "Overview of VHDL 92 standardization process", Proc. Euro-VHDL'91, Stockholm, September 1991, pp. 86-88.

[7]   VHDL-A Design Objective Document V2.1, IEEE 1076.1 Working Group, November 1994.

[8]   F. J. Rammig, "System level design", in Fundamentals and Standards in Hardware Description Languages, J.P. Mermet (ed.), Kluwer: Boston, MA, 1993, pp. 109-151.

[9]   L. W. Nagel, SPICE2, A Computer Program to Simulate Semiconductor Devices, ERL Memorandum ERL-M520, University of California, Berkeley, May 1975.

[10]   R. C. Rosenberg and D.C. Karnopp, Introduction to Physical System Dynamics, McGraw-Hill: New York, NY, 1983.

[11]   D. O. Pederson, "A historical review of circuit simulation", IEEE Trans. on Circuits and Systems, vol. CAS-31, no. 1, pp. 103-111, January 1984.

[12]   J. Vlach and K. Singhal, Computer Methods for Circuit Analysis and Design, Van Nostrand Reinhold: New York, NY, 1983.

[13]   A. Vachoux and K. Nolan, "Analog and mixed-level simulation with implications to VHDL", in Fundamentals and Standards in Hardware Description Languages, J.P. Mermet (ed.), Kluwer: Boston, MA, 1993, pp. 281-330.

[14]   A. Dewey and A. J. de Geus, "VHDL: Towards a unified view of design", IEEE Design & Test of Computers, vol. 9, no. 2, pp. 8-17, June 1992.

[15]   H. El Tahawy, D. Rodriguez, S. Garcia-Sabiro, J.J. Mayol, "VHDeLDO: A new mixed mode simulation", in Proc. IEEE Euro-DAC/Euro-VHDL'93, Sept. 1993, pp. 546-551.

[16]   R. Comerford, "Mechatronics fills the breach", IEEE SPECTRUM, pp. 46-49, vol. 31, no. 8, Aug. 1994.

[17]   A. R. Hefner and D. L. Blackburn, "Simulating the dynamic electro-thermal behavior of power electronic circuits and systems", in IEEE Workshop on Computers in Power Electronics, University of California, Berkeley, Aug. 9-11, 1992, IEEE Power Electronics Society, pp. 143-151.

[18]   H. A. Mantooth and M. Fiegenbaum, Modeling with an Analog Hardware Description Language, Kluwer Academic Publishers, 1994.

[19]   I. E. Getreu, "Behavioral modeling of analog blocks using the SABER simulator", in Proc. of Midwest Symposium on Circuits and Systems, Aug. 1989.

[20]   B. R. Stanisic and M. W. Brown, "VHDL modeling for analog-digital hardware designs", in Proc. IEEE International Conference on Computer-Aided Design, Nov. 1989, pp. 184-187.

[21]   G. Casinovi and A. L. Sangiovanni-Vincentelli, "A macromodeling algorithm for analog circuits", IEEE Trans. on CAD, vol. 10, no. 2, pp. 150-160, February 1991.

[22]   J. A. Connely and P. Choi, Macromodeling with SPICE, Prentice Hall: Englewood Cliffs, NJ, 1992.

[23]   D. L. Barton and D. D. Dunlop, "An introduction to MHDL", in 1993 IEEE MTT-S International Microwave Symposium Digest, IEEE MTT-S, June 1993.

[24]   P. N. Hilfinger, "Dimensional analysis in Ada", pp. 159-164 in Abstraction Mechanisms and Language Design, The MIT Press, 1983.

[25]   Proposal for Standard Math Package for VHDL, Version 0.9, Sept. 30, 1994, IEEE 1076.2 Working Group.

[26]   R. A. Saleh and A. R. Newton, Mixed-Mode Simulation, Kluwer: Boston, MA; 1990.

2

MODELING IN VHDL-A: DEVICES, NETWORKS AND SYSTEMS[1]

S. Peter Liebmann

Compass Design Automation 5457 Twin Knolls Road, Columbia, MD 21045, USA

ABSTRACT

With the advent of small channel semiconductor devices, highly dependent on proprietary processing technologies, there exists a need for flexible device modeling in a circuit simulator. One approach, taken by SPICE and SPICE type simulators, is to provide a parameterized empirical model, where the physical characteristics of the device are "fit" to the model. Such an approach is often limiting in that much of the physics of the device is hidden in the description, thus losing any predictability of behavior due to variations within a process. Alternatively, one can incorporate a physical model into the simulator, a task which usually requires a level of circuit simulation expertise most designers neither have nor desire. An alternative to such descriptive efforts is provided by an analog extension to VHDL, or VHDL-A. In this work, we illustrate how easy it is to model in VHDL-A since it allows a user to create and simulate a description of physical systems either on a very detailed level as for transistors, or on a more abstract system level. We show that VHDL-A modeling is no more difficult than writing down the equations which physically describe a system. As an example, we present a VHDL-A description of the small geometry MOSFET model with charge conservation of Yang and Chatterjee, since it not only contains physical

---

[1] This work was supported by the Rome Laboratory of the Air Force Material Command under contract # F30602-93-C-0209.

31

J.-M. Bergé et al. (eds.), Modeling in Analog Design, 31–45.
© 1995 *Kluwer Academic Publishers. Printed in the Netherlands.*

information inherent in most MOSFET models, but it also demonstrates the ease of modeling a complicated system in VHDL-A once a set of equations are derived. We also present two further examples to illustrate solutions of differential equations (a mixed signal DC motor) and idealized system macro-modeling (a function generator).

2.1.   INTRODUCTION

At the time of writing this manuscript, VHDL-A, the analog extension to VHDL, is undergoing an IEEE effort at standardization. However, enough of the concepts have already been developed and agreed upon to allow one to write many analog models in a style which, except for some possible syntactical changes, should appear almost identical to the official version of VHDL-A. While the scope of VHDL-A includes mixed analog/digital modeling, and modeling in both the time and frequency domain, in this work we explore the strength of this modeling language as related to three major areas in time domain analog design and simulation: device modeling, network design and system design.

To attain our goal, we present three examples which typify techniques used in the modeling of analog systems. Since the purpose here is to show how closely a VHDL-A model reflects the physical description, we present only enough of the language and the physics of the system to model it in VHDL-A and do not discuss the finer points.

The first example, relevant to IC designers, deals with the simulation of circuits made up of primitive devices composed of resistors, capacitors and sources controlled by voltages and/or currents. The descriptions of these devices are mathematical equations which characterize, to varying degrees of accuracy, physical properties such as a current as a function of a voltage. Most circuit simulators, like SPICE[1], have built in parameterized device models to describe everything from idealized resistors to sub-micron transistors. Designers, in turn, take various combinations of these devices and adjust the model parameters to provide a realistic representation of their system. However, with present advances in device technology, new devices are being developed with physical behavior which may not conform to the models provided by a circuit simulator. Changing models in a circuit simulator is often a very difficult and time consuming task.

VHDL-A provides a powerful tool to describe a new device in a simple straightforward manner. Once the equations which characterize a device are derived, a VHDL-A model can be developed directly from these equations and used by the circuit designers in their simulation.

In the first example, we present a simplified version of a device model developed by Yang and Chatterjee [2,3] to describe a small geometry MOS transistor with charge conservation [2] -[4]. We chose this model since a) it exemplifies how a VHDL-A description can be easily generated from a complicated set of physical and mathematical equations and b) it covers specific modeling techniques applied to semiconductor devices. We also show how a network made up of these devices can be described.

In the second example, we describe an analog network for an electro-mechanical system. We model a DC motor since it is a good example of an ordinary differential equation which, in this case, represents the mechanical part, interacting with a network which represents the electrical part.

Finally we show how macro-modeling is accomplished with VHDL-A. Macro-modeling is the technique of describing a complicated system by either its behavior or an idealized equivalent circuit, rather than by an actual circuit. Here we model a function generator with behavior described mathematically by a very simple transfer function.

In Section 2.2, we present an overview of VHDL-A and describe enough of the analog specific concepts and constructs so that one may understand the examples which follow. Section 2.3 is the mathematical description of the MOSFET which is modeled in VHDL-A in Section 2.4.1. Section 2.4.2 shows a particular modeling style for describing a circuit using the VHDL-A model in a manner similar to generating a SPICE netlist [1]. Section 2.5.1 contains the mathematical and VHDL-A description of the DC motor, and Section 2.5.2 presents the description of and code for a function generator. We present a summary and conclusion in Section 2.6.

2.2.   OVERVIEW OF VHDL-A

In this section we introduce enough information about VHDL-A so that one may read and understand the examples presented in this paper. While some of the concepts and constructs of VHDL-A are identical to standard VHDL, other constructs are needed mainly because analog simulation requires the solution of simultaneous differential equations while digital simulation does not.

The structural descriptions of analog systems are basically the same as digital ones, so that VHDL structuring mechanism [5] can be used to describe an analog circuit. However, analog systems differ mainly from digital systems in their behavioral descriptions. First of all, while digital waveforms are defined only at discrete times, analog waveforms are continuous with continuous time derivatives. Secondly, digital waveforms can be generated by an event driven simulator while analog waveforms are solutions to coupled algebraic - differential - integral equations. Analog systems are also specifically constrained by laws of conservation described by Kirchoff's current and voltage laws (KCL and KVL). Such differences necessitate the creations of new objects, types and constructs.

The first new object created for VHDL-A is called **quantity**. **Quantity**s are continuous in time and have time derivatives (**ddt**) and integrals (**integ**). They are the solutions to the coupled mathematical equations which describe an analog system. In other words, **quantity**s describe analog waveforms analogous to **signal**s describing digital waveforms. Since analog solutions are either real or complex depending on the type of analysis performed (real for transient, complex for small signal AC), a new class of type called analog is introduced which is a polymorphic type. For the purpose of the examples given in this text, we can consider an analog type as a double precision real number called float (we need high precision for accuracy in solving the analog equations).

In order to introduce the concept of conservation at analog connection points, a new object called **node** was created. **Node**s are characterized by two continuous waveforms with continuous derivatives called **through** and **across.** At any given time, the sum of the values of all **through** waveforms associated with a **node** is zero, while the values of all **across** waveforms at that **node** are equal. This rule governing **through** and **across** waveforms is a generalization of Kirchoff's laws to disciplines other than electrical (**through** waveforms are the currents and **across** the voltages of an electrical system).

The **through** and **across** waveforms of a node are solutions to the mathematical equations of many physical conservative systems [6]. In order to distinguish between connection points of different physical systems, a **node** belongs to a new class type called **nature.** A **nature** is made up of **through** and **across** waveforms and a reference node. Although the syntax of **nature** has not been defined at the time of writing, we will use the following:

```
type electrical is nature
     through i;
     across v;
end electrical;
```

The values of the **through** and **across** waveforms may be accessed once a **node** is declared. For example, if one declares a node "n1" one may write n1.i to be the current flowing into the node, and n1.v the voltage at the node. In this case, both are in reference to ground. To describe the current flowing from one node to another, we introduce a shorthand notation, [n1,n2].i which implies that if [n1,n2].i = x, then n1.i = -x and n2.i=x. With this notation for current, we define a branch between n1 and n2. Also, voltage between two nodes denoted as [n1,n2].v describes the voltage difference n1.v-n2.v.

Communication between different structures is accomplished through **coupling**s for **quantity**s and **pin**s for **node**s in a manner similar to **port**s being the communication mechanism of **signal**s.

Finally, in order to create the simultaneous mathematical equations which describe an analog system, we define a **relation** block. In this block, we may specify our equations either explicitly or implicitly. An example of an explicit description is current written as a function of voltage which defines a branch. Alternatively, implicit equations can be a system of 'n' equations with 'n' unknowns. Implicit equations are described in a **procedural** as in our MOSFET model, and explicit equations are in an **equation** as for the DC motor.

2.3.　　A MOSFET MODEL

To model a MOSFET for simulation, one first sets up an equivalent circuit [1] with four terminals, and then derives the current - voltage relationships which are solutions to Poisson's equations [2,7] applied to models which describe the branches of the circuit. A typical equivalent circuit is a current source between the source and drain, two reverse biased diodes between the drain and the substrate and the source and the substrate, and capacitors defined by the various MOSFET charges (we ignore subthreshold currents, contact resistors and other parasitics for simplicity - they are quite easily modeled once the equations are derived). The charges in a MOSFET, in turn, are caused by the PN junctions of the diodes, the overlap of the gate with the source and drain (which we will ignore, again for simplicity), and a charge related to charge conservation of a four terminal device [2] - [4] which is described by:

$$(1) \qquad Qg + Qs + Qd + Qb = 0$$

Here Qg, Qs, Qd and Qb stand for gate, source, drain and bulk (or substrate) charge. Equation 1 may be satisfied by three capacitors in parallel between the source, drain and gate and the substrate.

The system described above is represented by the equivalent circuit in Figure 1.

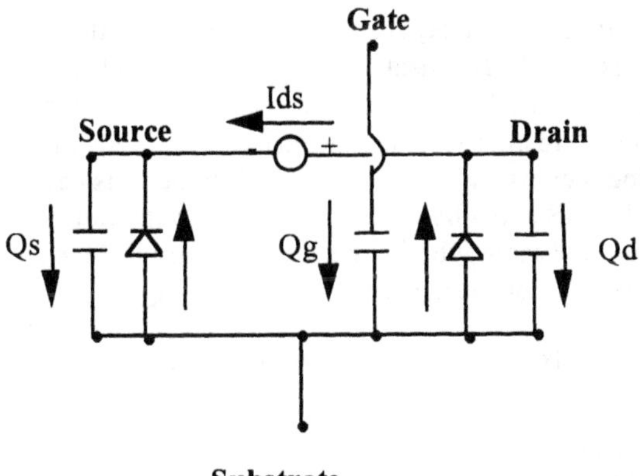

Figure 1:　The MOSFET Model

Here, the arrows represent the direction of positive current flow through the various branches of the device. The equations which follow are the currents through these branches written as a function of the three controlling voltages (Vds, Vbs, and Vgs).

2.3.1.   The DC Part

The DC part of the diode is given by:

$$(2) \quad Idiode = Is * (\exp(V/v_t) - 1)$$

where V is either Vbs or Vbd (the voltage at the bulk minus the voltage at the source or drain) and v_t is the thermal velocity. Is is the saturation current. The source - drain current equations are derived in this model by Yang and Chatterjee[2], and are given below:

2.3.1.1.   The Threshold and Saturation Voltages

The threshold voltage, Vth for a MOSFET is

$$(3a) \quad Vth = Vto + body * sqrt(2\ \phi f - Vbs)\ \text{ for an n channel MOSFET}$$

and

$$(3a) \quad Vth = Vto - body * sqrt(2\ \phi f - Vbs)\ \text{ for a p channel MOSFET}$$

where 'body' is the body effect and Vto is the zero biased threshold voltage (this term is simplified from Vth given in Ref. 2).

The current from drain to source, Ids, is zero if Vgs $\leq$ Vth so we will only consider currents where Vgs > Vth. If Vdsat is the saturation voltage, we may write the two regions of operation as: $0 \leq$ Vds < Vdsat for the linear region, and Vds $\geq$ Vdsat ≥ 0 for the saturation region. Ids, as well as all currents defining the various branches, are symmetrical around Vds when Vds is negative. Symmetry is defined as follows: given a function "f" of Vds, Vgs and Vbs which defines a current entering or leaving the source or drain for Vds ≥ 0, the same function defines the current if the source and the drain are switched when Vds < 0. For example, if

$$(4a) \quad Ids = f(Vds, Vgs, Vbs) \text{ for } Vds \geq 0$$

then

$$(4b) \quad Isd = f(Vsd, Vgd, Vbd) \text{ for } Vds < 0.$$

An algorithm reflecting this behavior is easily written in VHDL-A code and is given in Section 2.4.1. The saturation voltage given by:

$$(5) \quad Vdsat = (Vgs - Vth) / \alpha_x$$

where

$$(6) \quad \alpha_x = \alpha + \gamma * (Vgs - Vth)$$

In our case, the parameters $2\ \phi f$, body, α and γ are set by the users of the model.

2.3.1.2. The Current Equations

Ids or the source to drain current is given by:

$$(7a) \quad Ids = \alpha_x * \beta * (2(Vgs - Vth) * Vds) - \alpha_x * Vds^2$$
for the linear region

$$(7b) \quad Ids = \beta * (Vgs - Vth)^2 * (1 + \lambda (\alpha / \alpha_x)^2) * (Vds - Vdsat)$$
for the saturation region

Here,

$$(8) \quad \beta = Kp * W / L$$

where Kp and λ are parameters, and W and L are the channel length and width for the device of interest.

2.3.2. The Charge Model

For the sake of simplicity, we assume that the diode has a constant capacitance giving a charge

$$(9) \quad Qdiode = Cdiode * V$$

The other MOSFET charges are given in ref. 2 and need not be presented here since they are repeated in the VHDL-A code of Section 2.4 in much the same manner equations 2 - 8 are repeated. Like the MOSFET, we have three regions of operation, the linear, the saturation and the accumulation. We also introduce a new parameter, t_{ox} which is the oxide thickness. We can write the oxide capacitance as

$$(10) \quad C_{ox} = \varepsilon_{ox} / t_{ox}$$

for ε_{ox} the dielectric constant.

Once we know the charges, we know the branch currents:

$$(11) \quad I = ddt(Q)$$

2.4.   THE VHDL-A CODE FOR A MOSFET

Using the discussions of Sections 2.2 and 2.3, we can now write the VHDL-A code. From Figure 1, we see that we have four across variables, or currents; one between the source and drain, and three others between the gate, drain and source, and the substrate. These currents are described by the equations in Section 2.3 and ref 2. Parallel branch currents contribute to the total current at a node and are added together in accordance to the current flow in Figure 1. For example,

$$(12) \quad [drain, substrate].i = ddt(Qd) - Idiode + ddt(Qdiode)$$

In Section 2.4.1, we write the model for an NMOS and a PMOS. The PMOS equations are identical to the NMOS equations if the negatives of the voltage are used and the sign of the resulting current is reversed (a trick used in SPICE). In Section 2.4.2, we show an efficient coding style which may be used to instantiate MOSFETs described by the model.

In order to distinguish between NMOS and PMOS we define an enumeration type "mos"

```
type mos is (nmos,pmos);
```

which may be part of an analog transistor package.

In the following VHDL-A code, we assume certain variables like the thermal velocity (vt) and dielectric constant (ep) are given and other variables will have names similar to the description in Section 2.3. We also refer to the textural descriptions in the VHDL-A comments whenever possible.

2.4.1.   The "entity - architecture" for the Model

```
-- to model the Yang - Chatterjee MOSFET

entity y_c_mosfet is
    -- default model parameters
    generic(mos_type : mos := nmos; L, W : float := 100.0e-6;
            kp : float := 1.0e-5; phi: float := .8;
            alpha : float := 1.0; tox : float := 1.0e-9;
            gamma, body, vto, lambda, is, cdiode : float := 0.0);
    pin (nd, ng, ns, nb : electrical);
end y_c_mosfet;

architecture mos of y_c_model is
  begin
    relation
          -- voltage related
          variable n_type, vds, vgs, vbs, vth, vdsat: float;
          -- DC current related
          variable body, phil, alphax, beta, ids : float
          variable isdiode, iddiode : float;
          variable cox, vgst : float;
          -- to account for the symmetry defined in Section 2.3.1.
          variable switch: boolean;
          -- Charges
          quantity qsdiode, qddiode : analog;
          quantity qgate, qsource, qdrain analog;
```

```
procedural
-- set controlling voltages - see description of NMOS and PMOS
-- types given above.
    switch := false;
    n_type := 1.0;
    if type = nmos then
        n_type = -1.0;
    end if;
    vds := n_type * [nd,ns].v;
    if (vds < 0) then
        -- to take care of symmetry around vds see equations 4.
        -- Also, the description of NMOS and PMOS types given in
        -- the introduction to this section.
        switch := true;
        vds := -vds;
        vgs := n_type * [ng,nd].v;
        vbs := n_type * [nb,nd],v;
    else
        vgs := n_type * [ng,ns].v;
        vbs := n_type * [nb,ns].v;
    end if;

    -- threshold voltage - see equations 3
    phi1 := 0.0;                    -- Don't take the sqrt of a
    if (phi - vbs) > 0 then         -- negative number
        phi1 := sqrt(phi - vbs);
    end if;
    vth := n_type * vto + body * phi1;

    -- saturation voltage - see equation 5
    alphax := alpha + gamma * (vgs - vth);
    vdsat := n_type * (vgs - vth) / alphax;

    -- branch currents and charge: the charges are directly from
    -- ref.2, and the currents are from Section 2.3 of this work.
    cox := L * W * ep / tox;
    beta := kp * L / W;
    vgst := vgs - vth;

    -- accumulation region
    if( vgst <= 0.0) then
      ids := 0.0;
      qgate := cox * (vgs - vfb - vbs);
      qdrain := 0.0;
      qsource := 0.0;

    -- linear region - see equation 7a and reference 2
    elsif (vds <= vdsat) then
      ids := alphax * beta * (2.0 * (vgst) * vds)
            - alphax * vds * vds;
      qgate := cox *(vgs - vfb - phi - .5 * vds + alphax *
            vds **2 /(12 * (vgst - alphax * .5 *vds)) );
      qdrain := - cox * (.5 * vgst - .75 * alphax * vds +
        (alphax * vds) **2 / (8* (vgst - alphax * .5 *vds)) );
      qsource := - cox * (.5 * vgst - .25 * alphax * vds +
        (alphax * vds) **2 / (24* (vgst - alphax * .5 *vds)) );
```

```
                -- saturation region - see equation 7b and reference 2
                else
                  ids := beta * vgst**2 * (1 + lambda * (alpha/alphax)**2)
                       * (vds - vdsat)
                  qgate := cox * (vgs - vfb - phi - vgst/(3 * alphax) );
                  qdrain := 0.0;
                  qsource := -2 /3 * cox * vgst;
                end if;

                -- diode currents and charge - see equation 2
                isdiode := is * ( exp(vbs/vt) -1.0);
                qsdiode := cdiode * vbs;
                iddiode := is * ( exp((vbs - vds)/vt) -1.0);
                qddiode := cdiode * (vbs - vds);

                -- define the relative currents of Figure 1 - note how
                -- symmetry and the type of MOSFET is taken into account:
                [ng,nb].i := n_type * ddt(qgate);

                -- for symmetry definition see equation 4. Also, equation 12.
                if (not switch) then              -- vds is positive
                    [nd,ns].i :=  n_type * (ids);
                    [nd,nb].i :=  n_type * (ddt(qdrain) -
                                  (iddiode + ddt(qddiode)) );
                    [ns,nb].i :=  n_type * (ddt(qsource) -
                                  (isdiode + ddt(qsdiode)) );
                else                              -- vds is negative
                    [ns,nd].i :=  n_type * (ids);
                    [ns,nb].i :=  n_type * (ddt(qdrain) -
                                  (iddiode + ddt(qddiode)) );
                    [nd,nb].i :=  n_type * (ddt(qsource) -
                                  (isdiode + ddt(qsdiode)) );
                end if;
        end relation;
end mos;
```

2.4.2. Circuit Design with the MOSFET Model

One problem with designing a circuit using models like the one above is that in spite of
the fact that the model has many generic parameters, most component instances will, in
practice, use one of a small number of parameter sets. A typical case would be in
CMOS design where there would be one set of parameters for the n channel and another
for the p channel device. In the model described above, instances of the device would
only reflect changes in the channel length and channel width while all other model
parameters remain unchanged. Although in this work we described a model with
relatively few parameters, in many cases one may have up to one hundred or more
parameters with only about six changing for each device, as in the BSIM model of
SPICE3 [8]. It would be unreasonable to require a designer to repeat all the parameters
for each instance.

We demonstrate a modeling style here which addresses this problem and simplifies the
design. In our example below, a set of model parameters are written only once in an
entity - architecture pair. That entity may be referenced from different design units

which, in this case, represent an analog circuit description on the transistor level. Below is a code to demonstrate a typical design of an inverter using the small channel model of the last section. We first define two sets of parameters, one for NMOS and one for CMOS, and then instantiate the device.

```
-- set the model parameters for the NMOS device
entity nmos is
     generic(L, W : float);
     pin(nd, ng, ns, nb : electrical);
end nmos;

architecture model1 of nmos is
begin
  a1: entity y_c_model
     generic map(nmos, L=>L, W=>W, kp=>1.0e-4, phi=>.7, alpha=>1.2,
                 gamma=>.08, body=>.66, vto=>.8, lambda=>.07,
                 is=>1.0e-14, cdiode = 1.0e-12);
     pin map(nd, ng, ns, nb);
end model1;

-- set model parameters for the PMOS device
entity pmos is
     generic(L, W : float);
     pin(nd, ng, ns, nb : electrical);
end pmos;

architecture model1 of nmos is
begin
  a1: entity y_c_model
     generic map(pmos, L=>L, W=>W, kp=>.8e-4, phi=>.6 alpha=>1.01,
                 gamma=>.1, body=>.7, vto=>-.8, lambda=>.1,
                 is=>1.0e-14, cdiode = 1.0e-12);
     pin map(nd, ng, ns, nb);
end model1;

-- CMOS inverter
entity inverter is
     pin(in, out : electrical);
end inverter;

architecture cmos of inverter is
begin
-- assume vdd and gnd are global nodes defined somewhere
     m1: entity nmos
             pin map(nd=>out, ng=>in, ns=>gnd, nb=>gnd);
             generic map(L=>1.0e-6, W=>5.0e-6);
     m2: entity pmos
             pin map(nd=>out, ng=>in, ns=>vdd, nb=>vdd);
             generic map(L=>2.0e-6, W=>10.0e-6);
end cmos;
```

As we see in the above example, we may use the direct instantiation mechanism of VHDL'93 [5] to build a circuit with many transistors characterized by a large set of parameters by only specifying a few of the parameters per instance. In fact, this style of modeling is very much like modeling in SPICE with a netlist [1].

In a SPICE netlist, the entities nmos and pmos are replaced by ".MODEL". The components m1 and m2 are very similar to SPICE components in that "nmos" and "pmos" are the model reference, the pin map is the node list and the generic map elements are the device parameters.

2.5.   DIFFERENTIAL EQUATIONS AND MACRO MODELS IN VHDL-A

In the previous section, we showed how a device can be modeled once the mathematical equations describing the device are derived. In this section, we describe two systems, one where we write the mathematical equations down directly as a differential equation, and one where we use a simple mathematical function to describe the behavior of a system where the circuitry may not be known.

2.5.1.   A DC Motor

A DC motor consists of a mechanical rotational system, where an applied torque equals the sum of the reaction torques [6], and an electrical system. The applied torque results from a current induced by the electrical part. The angular velocity in the mechanical part, in turn, induces a voltage in the electrical part, thus forming a feedback loop.

The electrical part can be described by the circuit if Figure 2:

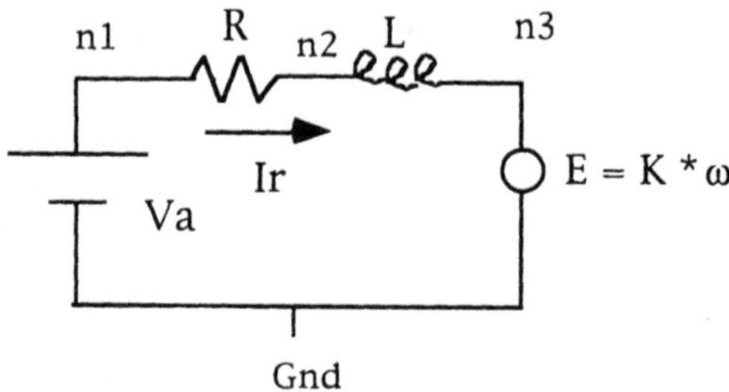

Figure 2:   The Electrical Part of a DC Motor

In Figure 2, K is the induced voltage coefficient and ω is the angular velocity.

If we let Kt be the applied torque coefficient, the applied torque is the product of Kt and the current through the circuit, which by KCL is equal to the current through R, or Ir. The resulting torque equation is given by the differential-integral-algebraic equation:

$$(13)\quad Kt * Ir(t) = J * ddt(\omega(t)) + B * \omega(t) + Ks * \int \omega(t') \, dt'$$

Here, J is the moment of inertia, B is the damping factor and Ks is the stiffness coefficient. t and t' are times.

We can now take the above description, and model it in VHDL-A as follows:

```
entity dcmotor is
   pin(n1 : electrical);
   generic(J, B, Ks, Kt, R, L, K : float);
end dcmotor;

architecture electro_mechanical of dcmotor is
     node n2, n3 : electrical;
begin
  relation
     quantity omega : analog;
     procedural
       [n1, n2].i := [n1, n2].v / R;              -- the resistor branch
       [n2, n3].i := 1 / L * integ([n1,n2].v);   -- the inductor branch
     equation(n3.v, omega)
           -- the induced voltage in the electrical part:
           n3.v == K * omega;
           -- the torque equations:
           Kt * [n1,n2].i == J * ddt(omega) + B * omega +
                       Ks * integ(omega);
  end relation;
end electro_mechanical;
```

As in the case of the MOS model, there is a direct correspondence between the behavioral or mathematical description of the system and the VHDL-A code. Here, we described the system as branch equations interacting with a differential equation. The schematic is easily described in the **procedural** part and the differential equation and feedback in the **equation** part.

It should be noted that this system may also be modeled entirely with branch current relations [6] with omega being the across variable and the torque being the through variable. We chose the above method, however, to illustrate how different modeling styles are supported in VHDL-A.

2.5.2.　A Function Generator

While in the previous example we modeled the circuit part of the system with primitive analog elements, in this example we model the behavior of a circuit with a simple mathematical function when no circuit description is given. We wish to describe a system with an input and output which can have any combination of the following:

(14a)　OUT = absolute value of IN if a rectify option is set.

and/or

(14b)　OUT = IN+ offset voltage

and/or

(14c)　OUT is clamped by some clamping voltage

In our example, we will consider a sine wave as the input which will produce a "reshaped" sine wave as an output. The output can be used as an external stimulus.

```
entity function_gen is
    generic(input_freq : float := 1.0e6);
    pin(pos, neg, offset, clamp, rect : electrical);
end function_gen;

architecture sin_gen of function_gen is
begin
   relation
     quantity input, output : analog;
     procedural
         -- set the input as a function of analog time (ANOW)
         input := sin (2 * pi * ANOW * input_freq);
         -- rectify the output and offset it
         if rect.v < 0.0 then
             output := input + offset.v;
         else
             output := abs(input) + offset.v;
         end if;
         -- clamp the output
         if output > clamp.v then
             output := clamp.v;
         elsif output < -clamp.v then
             output := -clamp.v;
         end if;
         -- set the voltage difference to the result
         -- this difference may be used as a stimulus to a circuit
         [pos,neg].v := output;
     end relation;
end sin_gen;
```

Again we see how the VHDL-A code directly reflects a behavioral description of an analog device, in this case on an abstract, rather than circuit level.

2.6. CONCLUSION

We presented in this work three different types of analog systems which arise in analog circuit design and simulation. In the first case, the solution to a differential equation describing a MOSFET was used to generate a closed set of equations which describe the device. In the second case, a differential equation describing a system was stated and related to a circuit through a feedback loop. Finally, we described the desired idealistic behavior of a system. We showed that VHDL-A codes representing the systems have one thing in common: once a description of a system is derived and presented either by a mathematical equation, an algorithmic description or a schematic, the VHDL-A code can be written in a straightforward manner directly from that description. Although software tools which fully simulate this language have not been developed at the time of writing, many tools have been developed which have produced fruitful results when applied to similar systems described in other languages. We feel that VHDL-A is going to be a major factor in analog design due to its ease of use and the fact that the systems described in this language can be simulated.

2.6.1.   Acknowledgment

I would like to thank Ken Bakalar at Compass for his invaluable suggestions especially about matters pertaining to VHDL.

REFERENCES

[1]   L. W. Nagel, "SPICE2: A Computer Program to Simulate Semiconductor Circuits", ERL Memo ERL-M520, University of California, Berkeley, May 1975.

[2]   P. Yang and P. K. Chatterjee, "SPICE Modeling for Small Geometry MOSFET Circuits", IEEE Trans. on Comp. Aided Des. of Int. Cir. and Sys., vol. CAD-1, no. 4, p. 169, Oct. 1982.

[3]   P. Yang, B. D. Epler and P. K. Chatterjee, "An Investigation of the Charge Conservation Problem for MOSFET Circuit Simulation", IEEE J. Solid-State Circuits, vol. SC-18, no. 1, p. 128, Feb. 1983.

[4]   D. E. Ward and R. W. Dutton, " A Charge-Oriented Model for MOS Transient Capacitances", IEEE J. Solid-State Circuits, vol. SC-13, p. 703, Oct. 1978.

[5]   IEEE Standard VHDL Language Reference Manual, 1076-1993.

[6]   D'Azzo and Houpis, "Feedback Control for Analysis and Synthesis", 2nd, McGraw-Hill, 1966.

[7]   For example, see A. S. Grove, "Physics and Technology of Semiconductor Devices", John Wiley & Sons, 1967.

[8]   B. J. Sheu, D. L. Scharfetter and P. K. Ko, "SPICE2 Implementation of BSIM", ERL Memo ERL-M85/42, University of California, Berkeley, May 1985.

3

ANALOG MODELING
USING MHDL

David L. Rhodes

Army Research Laboratory, Electronics and Power Sources Directorate,
AMSRL-EP-MA, Fort Monmouth, NJ 07703-5601 USA

ABSTRACT

A thorough description of the evolving MIMIC Hardware Description Language (MHDL) is presented from the perspective of an analog modeler or designer. Analog simulation issues pertinent to MHDL description are discussed. This includes both transient and steady-state simulation perspectives. Several modeling examples demonstrate the concepts. Simulation issues related to these examples, as well as MHDL- simulator interactions in general, are also discussed.

3.1.   INTRODUCTION

The process of creating a mathematical description of an object is called modeling. Obviously, any particular object, or piece of hardware, has many aspects which may be modeled. For example, it has a physical embodiment, thermal and electro-magnetic characteristics, etc. These 'major' descriptive regimes are many times referred to as *views* (i.e. a "physical" or "electrical" view) and the description or definition of characteristics within a view is often called a *behavior* (i.e., "thermal" or "digital" behavior). Other than the case of a very abstract model of behavior, behaviors within

47

J.-M. Bergé et al. (eds.), Modeling in Analog Design, 47–92.
© *1995 Kluwer Academic Publishers. Printed in the Netherlands.*

these views are not independent; that is, interdependencies exist among the views. Also, within a single class of behavior, for example electrical, several different models may be applicable, each with perhaps a varied level of detail or where each is more suited to particular use (e.g., a device-, circuit- or system-level model). These are often referred to as *levels of abstraction.*

The notion of level of abstraction is critical to modeling. While inherent laws (e.g., Maxwell's and Newton's equations) govern all physical objects, it is obvious that not much progress can be made be attempting to apply such laws at the most fundamental level to complex objects. The notion of level of abstraction captures the intended amount of detail and the characteristics which are of importance at any particular time. For example, a "signal flow" model of an OPAMP, which ignores the conservative Kirchoff's Current/Voltage Laws (KCL and KVL) but provides a simple abstract model, may suffice for the purpose of generating an initial circuit topology. In other cases, a model which incorporates detailed effects or is based on the composition of constituent elements may be necessary or desired. Use of different models, even within a single view, is also based on the purpose or use the model serves. As examples, a signal flow model cannot be used to determine voltage/current relationships; a model well suited to simulation might not be particularly useful for testing purposes. In summary then, selection of models must be made based on what particular characteristics/features are of prime importance and how the model is to be used, that is -- what are the questions to be answered.

As the title implies, here we are primarily concerned with analog modeling. Some of the distinctive characteristics of analog (vs. digital) are: continuous valued range (non-quantized) and continuous valued domain (i.e., non-discrete time axis); handling of conservation (e.g., KCL, charge) rules; etc. All of this is in stark contrast to a digital description, where models have inputs and outputs (i.e., a directionality is implied/determinable) and there is no notion for conservation or support for equations. Even pins defined as "inout" are in either the "in" or "out" (or possibly high-Z) state at any particular time point. Moreover, in the digital regime only time domain simulation is generally considered. This regime is hence quite readily adapted to a state-based, procedural language definitional style. Even so, a language such as VHDL [VHDL-93], by virtue of being a computer language in itself, is adaptable and extendible to purposes other than what it was originally intended. For example, VHDL resolution functions were used in [Zhou-91] to perform some measure of analog simulation.

A Hardware Description Language (HDL) is a term used for a language capable of capturing features of a hardware object. Before examining the usefulness of the **MIMIC Hardware Description Language (MHDL)** for analog modeling and its application to simulation, an overview of MHDL, with emphasis on elements related to electrical component modeling will be given in the next section, called "Features of MHDL". Although this section discusses language features from the perspective of an analog engineer, the casual reader may prefer to skip this section. A discussion of analog modeling types is next contained in Section 3.3, "Model and Simulation Types". This is followed by the "Examples" section which shows MHDL modeling and extractions for a variety of electrical circuit/system classes. Consideration for how a language will be used is just as important as the language design itself; this is discussed in Section 3.5, "CAD/CAE Tool Interaction". Conclusions are then contained last in Section 3.6, "Concluding Remarks".

3.2. FEATURES OF MHDL

Before delving into MHDL's language design, some brief background comments are in order. The origin of MHDL sprung from the need to exchange microwave and analog models, and provide a common framework for computer-aided engineering (CAE) tools. Although "birds-of-a-feather" meetings concerning the possibility of an HDL for analog circuits and systems date back to at least 1984, MHDL began formally in 1988 with the gathering of requirements. The result from this effort was the "MHDL Requirements Document" [MRD-91]. Using this as a basis, the language design for MHDL began in 1991. MHDL is formally defined in its language reference manual [LRM-94] which represents it's fourth major iteration/document revision. Currently, MHDL language modifications are under the province of the IEEE SCC-30 and the description given here is of course based on the current definition. Another point to note is that, subject to scoping and implementation concerns, it is planned that MHDL conform to requirements not only as expressed in [MRD-91] but also to those determined by IEEE SCC-30 [SRD-94]. Currently, several organizations are developing software to demonstrate MHDL, however, these efforts are not sufficiently mature to allow illustration of actual results in this paper.

Before moving to examples, which serve to illustrate the use and application of MHDL in the regime of analog modeling and simulation, a basic understanding of MHDL is necessary. An HDL for analog modeling by its very nature must span two disciplines, namely electrical engineering and computer science. Any formally defined language, such as MHDL, must be well founded in terms of syntax and grammar, and most importantly in terms of (execution) semantics. While the character set, lexical groupings and syntax provide the structure/form of the language, semantics define the meaning of such. Most computer languages are procedural, in that they are defined in terms of a sequential execution of a set of statements (e.g., FORTRAN, C, and to an extent VHDL [VHDL-93]). On the other extreme are non-executable languages (e.g., EDIF) which provide data exchange capability but are not usually adapted to direct use in simulators or more generalized CAE backplanes.

Supporting the analog modeling (and simulation) regime poses some severe challenges to an HDL. First, behavior of basic analog elements does not allow determination of "directionality" at model interconnection points. For example, a resistor does not have an input or output, but two terminals for which only a voltage/current relationship must hold. Additionally, conservative laws at interconnections (e.g., KCL) must be supported. Furthermore, in general, modeling for multiple, interrelated domains (e.g., time, frequency, Laplace) must be supported and precise meaning (or error conditions) must be given to the interconnection of such. The HDL should also be capable of defining values in non-deterministic (statistical) terms. Finally, in addition to support beyond simulation, the HDL should be capable of supporting at least transient, small-signal (a.c.), and noise analyses.

MHDL tackles the challenges in analog modeling (and simulation) by exploiting a functional language technique. Specifically, MHDL is based on an executable form of lambda calculus (the calculus of functions). As will be seen, this approach naturally supports many areas needed for analog modeling (signals, equations, multiple views). The following subsections serve to reasonably completely define MHDL from the perspective of an analog modeler, but keep in mind that MHDL is indeed a computer language. Throughout the text, a `courier-font` is indicative of actual MHDL text or language keywords or phrases.

3.2.1.   Models, Packages, Structures, and Attributes

The outermost elements of an MHDL description are models and packages. Models in MHDL naturally correspond with hardware modules or components in a physical (or envisioned) system. Packages are used to facilitate the sharing of common information, constants, functions, etc. In MHDL there is one particular package, `Standard`, which has special status in that all models and other packages inherently make use of it. It defines constants and functions which have wide use (i.e., pi, mathematical functions, etc.) as well as other things to be discussed later. As explained above, there may be many ways to view a particular analog hardware element (e.g., connection terminals might be viewed in terms of voltage/current characteristics, or in terms of physical connector type and location) and there may be several levels of detail of interest (e.g., a signal-flow versus a conservative voltage/current model). To support this, MHDL allows multiple structures within each model. That is, each structure within a model might be representative of either various views or various levels of detail within a particular view.

Before completing the description of hierarchy, some knowledge of attributes must be given. Assignment of values to attributes is a basic mechanism for modeling and describing hardware in MHDL. The language statement is called an assignment or, equivalently, a property name definition. Although the term "assignment statement" is used, there is no notion of sequential execution of these (nor in MHDL at all). Such assignments can occur in any textual order within the description with identical meaning (the language evaluator is in fact constrained to evaluate these with respect to data-dependency but the user can place them in any order).

An essential constructive technique that must be supported is of course hierarchical arrangement of components, including passing "actual" values to "formal" arguments to particularize a general model (e.g., the resistance parameter of a resistor model is called a formal argument, the value it receives when the model is used is called the actual argument). As discussed above, particular views of a model are contained in structures, and it is therefore structures (in models) which are instanced within other structures to form a hierarchy. In MHDL, the instanced structure is called a component. The `definitions` clause is used to specify values for attributes in the structure being instanced. A very simple example serves to illustrate this, consider:

```
model resistor
structure electrical
  attribute r : Resistance;
  default   r = 100 'ohms';
  -- ... rest of definitions, connectors, etc.
end electrical;
end resistor;
--
-- now use it
--
model main
structure electrical
attribute r : Resistance;
  components
    R1 :: resistor.electrical definitions
          r = 45 'ohms';   end R1;
  end components;
end electrical;
end main;
```

In this case, a model called `resistor` contains a single structure called `electrical` which has an attribute called `r` (an actual argument). A default assignment statement (which is optional) provides value of 100 ohms for attribute `r` should no other value be assigned in the invocation. A value of 45 ohms (the formal argument) is however provided in when the `R1` component is created from the `resistor.electrical` structure (hierarchical names using the dot (.) notation are used throughout MHDL for `model.structure` and `structure.connector`, etc. forms). Note that both the type system (used when `r` is typed as a `Resistance`) and unit handling (since `r` takes on units of ohms) are discussed in subsequent sections, here the emphasis is on hierarchy and parametric models. Also note that the attribute `r` is viewed as being "attached to" the structure `electrical`, but as will be seen later attributes can be attached to many parts of a description.

Along with the "flow of information" along hierarchical lines, MHDL provides both multiple inheritance and information hiding. The `include` statement and inclusion list are used to refine or add information in models. Suppose that a (still fairly simple) thin-film resistor which depended on physical width and length dimensions were to be created. A model could be created which used the simple `resistor.electrical` as a component, but MHDL provides additional mechanisms for creating a derived model. Consider:

```
model tfr includes resistor;
structure electrical
  attribute w, 1 : Length; -- parameters
  r = some_function(w,1) 'ohms';
end electrical;
end tfr;
```

Here the (more specific) model `tfr` is *derived from* the general purpose `resistor` model without necessitating an additional layer of hierarchy. In order to support both the concepts of derived models and generalized information refinement, the rule is that all included or inherited definitions are combined in structures of the same name (the usefulness of this will be further demonstrated in Section 3.4.5). Since the structure in `tfr` is called `electrical`, the single definition for `r` will be combined with other definitions from `resistor.electrical`. In this case, the default assignment for `r` will of course be overridden and the value given by `some_function(w,1)` will be used. Of importance is that all other assignments (as well as equations, connectors, etc. to be discussed later) apply to `tfr.electrical`.

In addition to this "special" form of structural combining, the include keyword is also used to provide inheritance. In modeling, there of course arises the need to make use of functions, constants, type declarations, units, etc. Obviously, there exists a need to be able to share and reuse packages other than `Standard`. The include statement is used for this purpose:

```
model xyz
includes my_math_package, another_package;
--  ...
end xyz;
```

Note that the `Standard` package can be thought of as tacitly being included last, if it appears explicitly in an include list then this order can be changed accordingly. Conceptually, all attribute definitions, types, dimensions, structures, etc. can be thought to be textually included at the point of the include statement; however the actual language mechanism is more like the multiple inheritance of languages like C++. With the basic mechanisms for inheritance, hierarchical construction and derived models, the next major topic is connectors and their semantics.

3.2.2. Connectors and Connections

The need to support varying views of a "connector" exists in analog modeling. For example, a connector could be a physical connector (i.e., a coaxial connector or IC pin) or merely an abstract interaction point. The *typing* of connectors forms part of the support for dealing with this wide variety of connection interactions. Just as the there is a wide range of possible connector types, there is an equally wide range of desired interactions among connectors. As examples, it might be expected that KCL hold at the connection of electrical domain connectors, or that heat conservation applies for thermal connection flows, or that a user supplied convolution routine should be used at the connection of frequency and time domain models. The connection statement is used in MHDL to infer a connection between two or more connectors, and the issue as to the exact meaning of such is called *connection semantics*.

In recognition of this wide variety of connector types and connection semantics, MHDL provides only a basic connection semantic within the language proper but moves most of the other "expected" engineering laws and rules into executable code and packages. For example, KCL/KVL are contained in Standard and apply by default to connections of electrical type, but the user can provide and use, in say an included package, other connection rules. Note that, just as Standard is a package inherited by default, MHDL_Standard is a library of provided packages and models included with MHDL.

Currently MHDL connection semantics are strongly based on typing of connectors as well as on the names and types of attributes attached to connectors. Several MHDL "built-in" functions supply support for writing connection rules, these arc called *structural functions*. Only those structure functions which arise in subsequent discussions will be presented here, see [LRM-94] for a complete list and more precise definitions. The following structure functions operate over the entire design hierarchy:

connected_obj(con,prop) returns a list of all connectors connected to con which also have a property, prop, attached to the connector. This list provides the values for each of these as well. Properties are either signals (discussed below) or attributes. For example, this function can be used to find all other connectors attached to a particular connector which have an attribute "v" defined on them (along with the values for each "v").

structure_obj(struct,prop) is similar in that returns a list of all connectors in struct which have a property (signal or attribute), prop, attached to it (along with the values for prop).

The purpose of the following structure functions is to provide "local" functions related to connectors, that is those which apply only to connectors within a structure and those directly in components (within a single layer of hierarchy). Those that will be used later are:

child_str_obj(struct,prop) provides a list connectors which have an property, prop, and which are defined in components of struct.

sum_chistr_objects(struct,prop) provides the sum of all properties, prop, on connectors defined in components of struct. Note that this function is defined in terms of the child_str_obj function.

sum_parstr_objects(struct,prop) provides the sum of all properties, prop, on connectors defined directly in struct. Note that this function is defined in terms of the structure_obj function.

The resistor and main models first shown in Section 3.2.1 will now be shown with connectors and connections in place:

```
model resistor
structure electrical
  attribute r : Resistance;
  default   r = 100 'ohms';
  connectors p1, p2;
  p1.ctype = electric;
  p2.ctype = electric;
  -- ... still more definitions, etc.
end electrical;
end resistor;
--
-- now use it
--
model main
structure electrical
attribute r : Resistance;
  connectors m1, m2;
  components
    R1 :: resistor.electrical definitions
          r = 45 'ohms'; end R1;
  end components;
  Net1: connect m1, R1.p1;
  Net2: connect m2, R1.p2;
end electrical;
end main;
```

In the `resistor` model, two connectors, p1 and p2, are declared. Additionally, an attribute `ctype` is attached to each of these connectors and given a value "`electric`". As stated earlier, connector typing specifically via the attribute `ctype`, is used to apply standard MHDL connection semantics. Connectors (m1 and m2) are also defined in `main.electrical` and two connections (labelled `Net1` and `Net2`) are used which establish interconnects between m1 and `R1.p1` and between m2 and `R1.p2`.

3.2.3.   Special Names and Equations

Essential elements in the definition of analog models are the notions of time, frequency, spatial coordinates, etc. The special names, `t`, `f`, `s`, `Z`, `xdim`, `ydim`, `zdim` are defined in MHDL as time (seconds), frequency (Hertz), the Laplace domain variable (unitless), the Z domain variable (unitless), and the three dimensional physical coordinates (meters), respectively. Note that MHDL is *case sensitive* so that the capitalization of Z is significant. While these specially named attributes can be freely used in modeling, they are not allowed to be assigned to (they are always independent variables).

As mentioned above equations are necessary to capture analog models and behavior. Equations are directly supported in MHDL via `equation` statements. The name "equation" in MHDL is a bit of a misnomer in that any Boolean valued result, including inequalities or function results are allowed in equation statements, but they are typically used to express equalities. The semantic implied by MHDL equation statements is that the boolean result must be true (always). However, MHDL does not in itself provide for solution of the simultaneous expressions constructed.

The conclusion of the `resistor` model above along with a complete ideal capacitor model appears as:

```
model resistor
structure electrical
  attribute r : Resistance;
  default   r = 100 'ohms';
  connectors p1, p2;
  p1.ctype = electric;
  p2.ctype = electric;
  EQ1 : p1.i(t) == (p1.v(t) - p2.v(t)) / r;
  p2.i(t) = -p1.i(t);
end electrical;
end resistor;
--
model capacitor
structure electrical
  attribute c : Capacitance;
  connectors p1, p2;
  p1.ctype = electric;
  p2.ctype = electric;
  EQ1 : p1.i(t) == c*derivative(t)(p1.v(t) - p2.v(t));
  p2.i(t) = -p1.i(t);
end electrical;
end capacitor;
```

Note that no default (or other) value was given for c, implying that this attribute is required when the structure is instanced (used) as a component. This example illustrates many other important aspects in modeling. First, in order to be compatible with the standard header (Standard), current must be defined as positive when flowing into the model. Secondly, the connection semantics provided in Standard assume (for electrical connections) that the attributes v and i are attached to the connectors and also that they are defined as functions from (Time -> Potential) and (Time -> Current) respectively. We will return to the fact that attributes can be functions (i.e., functions are first-class in MHDL since they be assigned, modified, passed as arguments, etc.), but first will discuss other aspects which this example brings up (except units which are discussed later).

Looking at the capacitor model, the statement beginning with the EQ1: label is an MHDL equation. This equation, of course, expresses the current - voltage relationship for an ideal capacitor. This also makes use of one of the few non-LRM specified functions, namely the derivative. That is, in order to allow either symbolic or numeric solutions, the LRM defines only the mathematical meaning for the derivative, integral and some related functions but does not constrain their implementation. As explained previously, this same argument applies to equation statements as well. Thus, although MHDL enforces the mathematical intent for equations and derivatives/integrals, evaluation or use of these statements are subject to externally supplied tools (CAD tools, simulation in this case).

3.2.4.   Types and Classes

While a prime intent of any HDL is to be natural for the intended user, HDLs are indeed computer languages and are therefore subject to rules and theory governing them. Although the topics contained in this subsection might be viewed as more related to "computer language theory" than to analog modeling, these topics are critical to full understanding of MHDL and are actually of prime importance to the user as they provide the basis for many 'ease-of-use' features.

First, all computers use different instructions/hardware for processing of different types of operands. For example, integer and floating point addition on most computers utilizes different machine instructions as well as data in different forms. The form of the data is normally captured in an HDL via a *type* (e.g., an integer- or a floating point-typed value). Although untyped languages have been implemented, typed-languages are far more efficient since they can more closely match language operations and data types to those inherent to the computer. Therefore MHDL, along with almost all other programming/HDL languages, is indeed typed. MHDL provides both "built-in" types, like integer, floating point, arrays, lists, and also enables the user to create new types via the type statement. Note that although we have concluded that types are necessary via the tie to computer hardware, this fits with the "classic" view that a type is a named subset of values from the entire set of values computable or representable by a language.

In the case of low-level functions (i.e., integer addition) which directly correspond to computer execution, the 'linkage' between computer instructions and MHDL is via `primitive` functions, although any function (i.e., written in C or FORTRAN) can be used in MHDL via the primitive function capability. Hence, the addition of two integers or two floating point numbers is implemented via primitive functions. Since 'operators' like + are actually infix functions, MHDL provides both a traditional infix notation (3+4) or a function application (prefix) form where the operator is surrounded by parentheses (e.g., (+)(3,4)). Since the infix notation is merely a syntactic convenience, binary operations (and unary negation) are properly viewed as function applications.

Although the computer will need to execute different instructions for the addition of two integers versus two floating point numbers, at the language (MHDL) level the use of the same symbol (e.g., "+") is a requirement. For example, it is not acceptable to use "+" for integer addition and, say, "`float+`" for floating point operands. While languages like FORTRAN had some built-in *ad hoc* features for providing this capability (e.g., the + symbol could be used for integers and floating point), MHDL, as well as other modern languages, provides such capability via *polymorphic types* and *coercion* which allow extension of such features to the user level. Note that FORTRAN also provided "generic" functions (i.e., `sin`) which could take an argument of type floating point, double precision or complex, rather than force the user to use to non-generic (actual) functions (e.g., `SIN, DSIN, CSIN`). This is actually a simpler and less general technique called *overloading*.

MHDL formalizes operations (functions) on polymorphic types via the use of a *type class* (or just class) system. Simply put, a class is a named set of functions (technically called *methods* of the class, but they are just a list of functions). The `class` statement is used to list the names of functions (methods) belonging to the class and also declares argument lists of the these functions. In most cases, the argument list will contain polymorphic types; a polymorphic type is merely a composite of other not yet specified types. Note that (in general for any function as well as those which are `class` methods) arguments and results can be polymorphic, meaning that they may be any of a number of types. MHDL (class) `instances` are then used to create "type-specific" versions for specified types for all functions (methods) in the class (this process is also called *specialization*). Remember, it is the fact that each `instance` (specialization) must implement **all** of the functions in the class which enables coercion and polymorphic functions in a well-defined and user extendible manner. For example, if an "array" type is declared to be a member of the same class which contains (+), then by caveat it must provide an implementation of (+) for its type (an `instance`). So, by virtue of being a member of the class, it is known that all functions in the class must be implemented and hence the correct (+) function can be applied, in this case the "array" version.

To briefly illustrate this, the class `Group` is the class in MHDL which contains the (+) function, part of which appears as:

```
class Group (a) {
   (+),(-) :: a -> a -> a;
   -- ...
```

which says that the functions (+) and (−) belong to class `Group`, and that each of these functions has two arguments and results in a value of the same type. Note that first two "a"s in a `->` a `->` a are the arguments and the last one is the result value. More importantly, note that the "a" is a (polymorphic type) 'place holder' which will have a type substituted for it in, possibly multiple, `instance` statements. As explained above, the instance statement provides a specialization of the class. For example, the following:

```
instance Group(Int) {
   (+)  = primitive "PlusInt";
   (-)  = primitive "MinusInt";
   -- ...
```

is the MHDL text which provides the integer valued versions (the specializations) of addition and subtraction. As was noted at the beginning of this subsection, their implementations are as primitive functions. If the user desired to use (either the prefix or infix) + functions on another type, another instance statement will simply provide this. The actual functions could be either primitives or written in MHDL.

MHDL uses *type constructors* to enable the formation of values which would otherwise appear as either general purpose values or become ambiguous. For example, consider the type for a rectangular (2 dimensional) point in space which would likely be represented as a pair of two floating point values as:

```
type Point = Pt (Float, Float);
```

where `Pt` is the type constructor. A value of, say, (9.0, 6.5) *could* be interpreted as a value with type `Point` but could also be interpreted as *any* two element tuple value or any other type which is represented as a pair of floating point values. However the value `Pt (9.0, 6.5)` is unequivocally a value with type `Point`.

A final regard to types in MHDL is the mechanism for *subtyping*. A subtype is a type which contains a restricted set of some type, called the *supertype*. Subtyping is most often found in the context of restricted ranges in other languages. For example, an integer type which ranges only over 1 to 4 can be declared in VHDL. The first practical issue related to subtypes is that all functions defined for the supertype should automatically apply (be applicable to) to the subtype. Any function defined for the supertype obviously is applicable to the subtype and MHDL of course supports this. While other languages restrict the definition of subtypes to only that of ranges over the supertype, MHDL allows any boolean valued function to determine subtype membership. So for example a subtype of odd integers would be declared as:

```
type odd = Odd (Int) where { Odd (i) = is_odd(i) };
```

an attempted assignment of a non-odd number (as determined by the `is_odd` function) to an attribute with type `odd` would result in an error. Note that the term `Odd` is a type constructor which "uses" an `Int` to create its value.

3.2.5.   Coercion

While it is likely (and is the case in MHDL) that three different routines will be provided for "+" for the types `Int`, `Float` and `Complex`, what about functions involving combinations of these types? That is what about expressions like 1 + 4.3 where the arguments are mixed? Obviously, the use of the "+" operator in mixed-valued expressions with integer, floating point, or complex valued operands is also essential to ease-of-use and is in fact a requirement [MRD-91]. Although it might be conceivable to provide functions for all possibilities, this approach would suffer from a combinatorial explosion in the number of implementation functions (instances) needed. For example the `class` and `instance` phrases in the case of adding an integer and float would need to be modified to:

```
class Group(a,b) { -- NOT DONE THIS WAY:
  (+),(-) :: a -> b -> b;
  -- ...

instance Group(Integer,Float) {
  (+) =  primitive "PlusInteger_then_Float";
  -- ...
```

This would of course be even worse for functions with more than two arguments rendering such an approach intractable.

Rather than try to use the class system to achieve the goal of mixed typed operands (for all functions not only (+)), *coercion functions* can be used to convert arguments into forms compatible with type signatures. Related to the case above, MHDL contains the following:

```
coercion InttoFloat :: Int -> Float;
InttoFloat = primitive "InttoFloat";
```

which states that a coercion from the type Int to Float exists and also provides the implementation for it (as a primitive in this case).

Looking a bit into the operational mechanisms, at the time the expression 1+4.3 is encountered, a type instance for (+) is found which looks like instance Group(Float) which declares and defines (+) with a type signature Float -> Float -> Float. The coercion function from Int to Float is also known, so that the mechanism is to first apply the coercion function to the argument 1 to create a floating point 1.0 and then the floating point version of (+) is used (creating a floating point result). It is important to realize that this type analysis and function binding is done at "compile" time so that there is no "run-time searching". In fact, the importance of compile time (early) error/type checking was recognized in [MRD-91] and put forth as a requirement for MHDL's design. MHDL is, in fact, a *strongly* and *statically* typed checked language. The term "statically" refers to the ability to determine types at compile time, while the term "strongly" refers to the ability to detect illegally typed function arguments and attribute assignments.

Thus, the net effect of (user-extendable) polymorphic types and coercion, formalized using type classes, is to provide an efficient and adaptable expressions and functions while avoiding mandating explicit "type-casting."

3.2.6.   Functional Basis and Lazy Evaluation

MHDL also enjoys other benefits and techniques stemming from its functional programming heritage which are useful to analog modeling. The prime benefit is that functions themselves are "first-class" in that they can be manipulated as data, modified, passed as arguments, etc. For example, the voltages appearing at electrical connections and currents flowing into models are defined (using the default) as functions from (Time -> Potential) and (Time -> Current) respectively. Thus the actual physically occurring "event" is captured directly as a type in MHDL. Such values, that is ones which are really functions, are called *functionals*. Often occurring *physical signals* are also captured in MHDL's class system using functional types. For example, the class Per captures the essential aspects of periodic signals. It contains functions to return the (fundamental) frequency and phase of a given signal in the class Per and to define/create a signal in class Per with supplied characteristics.

Another significant advantage of the functional basis is that functionals can act as "in-line" function definitions. Consider the fragment:

```
attribute sqr : Float -> Float;
attribute a, b, c : Float;

sqr(x) = x*x;        -- OR sqr = \(x) -> x*x;
a = 5;
b = sqr(a);          -- b is 25
c = sqr(a+b);        -- c is 900
```

where the `sqr` utility function used to define b and c does not need to be defined separately. This is significant to both reduction of input code, modularity and even efficiency.

Although MHDL functions are executable (that is, can be evaluated), they are defined using more of a mathematical style rather than through the use of execution of a series of statements. The following fragment serves to illustrates some MHDL functions in combination with some user defined types and specializations (instances):

```
type Point = Pt (Float, Float);        -- a Point is a pair of Floats,
                                       -- representing x & y coords.

instance Group(Point) { -- specialize Group for Point types
    (+)=primitive "PlusRectang";
    (-)=primitive "MinusRectang";
    };-- use exactly same primitives as complex numbers in Rect. form

realp(x) = y where { (y,_) = x; };     -- first tuple element
imagp(x) = y where { (_,y) = x; };     -- second/last tuple element

distance :: Point -> Point -> Float;  -- function declaration optional
distance (point1, point2) = sqrt(xdif^2 + ydif^2)
        where{ xdif = realp(point1) - realp(point2);
               ydif = imagp(point1) - imagp(point2);};
```

In order to save space and keep the above example somewhat brief, the type `Point` was only declared to be part of the class `Group`. Function types use the "->" nomenclature, the type for the distance function (`distance :: Point -> Point -> Float`) should be read as `distance` takes a `Point` argument (first `Point`) and a `Point` argument (second `Point`) and returns a `Float` value. The underscore (_) in the definitions for `realp` and `imagp` represents the fact that this portion of the value is not needed; more precisely this represents a don't care situation in the *pattern matching* of arguments, a full treatment of which is beyond the scope of this paper. As a strong theory of *type inference* exists and is used in MHDL as well as in functional languages, the function declaration is optional in that argument and result types can be inferred automatically. Thus, function type declarations are not generally needed, but can be an useful aid to an MHDL compiler or used to restrict the function to more specific type(s).

Functions can also be created via a process called *currying*. Conceptually, currying creates a more specific function from a more general one by supplying one or more arguments to the general function. For example, suppose the `distance` function above was already available and a (single argument) function which determines the distance from the origin was desired. This can be created from the `distance` function as:

```
distance0 :: Point -> Float; -- function declaration

distance0(p) = distance(p, (0.0,0.0));

-- OR IN EQUIVALENT CURRIED FORM
distance0 = distance( , (0.0,0.0));
```

the last of which defines a `distance0` function curried from distance (note that the first argument is omitted but the second supplied). A high-degree of re-use can be obtained by using the style of writing general purpose functions and then using currying to obtain more specialized functions.

As may be evident, MHDL (and other functional languages) view input (programs) not as sequences of operations to perform but as a set of interdependent values to compute. While "conventional" programs, or models, are rather readily transliterated into MHDL, *loops* usually present the largest challenge. However, usually examination of the *purpose* of the loop renders the solution. Loops usually serve two purposes: (1) to affect the initialization of and/or operations on multiple dimensioned values (e.g., arrays, matrices); or (2) iterative computation of a result via updating. The MHDL approach for the first involves *list comprehensions* while a recursive function approach is generally used in the later. Some simple examples are:

```
attribute 1 : [Int];        -- type of "1" is list of Int
1 = [ 0 | i <- [0..100] ];  -- "1" is a list of 100 zeros

fac(0) = 1;
fac(n) = n*fac(n-1);      -- NOT { fac=1; for(i=1;i<n;i++) fac = i*fac; }
```

The "1" example shows a list comprehension for initializing a list of integers to zero while the factorial example shows recursive definition over the iterative one for the factorial function. The concept of introducing state variables for the purpose of directly supporting loops is a topic of recent discussion but is not part of MHDL currently.

One final important major topic related to the functional basis for MHDL is lazy evaluation. Lazy evaluation is one where values are not computed if and only if they are needed. For "low-level" functions, lazy evaluation comes into play more in the context of pattern matching, list and array handling, etc. For example, for the `realp` function defined above, if the following where used: `realp( (7,sqrt(100)+exp(7^3) )` the second argument never gets evaluated since it is not needed to arrive at the value 7. Lazy evaluation also plays an important role at higher levels, for example, a description might include electrical, thermal, and other views which may be either unrelated (mathematically), or partially or strongly interrelated. The lazy evaluation semantic will "automatically" only evaluate portions necessary to "answering" the "question" being posed.

3.2.7.   Types of Derivatives and Integrals

Since derivative and integral expressions are essential to analog modeling, a brief mention of their respective types is in order. MHDL, being a computer language, requires properly typed expressions. Therefore consider the types for the derivative and integral MHDL functions:

```
derivative(a)( a -> b )        has a type    a -> b / a
integral(a)( a -> b )          has a type    a -> b * a
integral(a)(c1,c2)( a -> b )   has a type    b * a
```

Remember that we are discussing the **types** of these expressions, not their **values**! For example, an expression which has the type (Time -> Capacitance), when derivated over t (Time), that is an expression like derivative(t)((1-t*2'Hz') 'pF') or in mathematical form with units $\frac{d}{dt}\left[(1-2t)pF\right]$, has a type Time -> Capacitance/Time since the resultant derivative provides the slope of the range type per domain type, for any domain point (note that range is Capacitance and domain is Time in the example). The second case listed above is the indefinite integral. As an example, consider now the integration of a (Time -> Capacitance) expression over t (Time) which appears like integral(t)((1-t*2'Hz') 'pF') in MHDL and as $\int \left[(1-2t)pF\right] dt$ in mathematical form. The resultant type of such an expression is (Time -> Capacitance*Time). Finally, the last line above shows a definite integral, in this case the resultant type is the product of the types (area). For example, a definite integral form for the previous expression (i.e., $\int_{0}^{5} \left[(1-2t)pF\right] dt$) results in an expression type Time*Capacitance that is, in definite integral form, the dependency on t is eliminated.

The case of the continuous Fourier integral is interesting. The Fourier integral is written as: $F(\omega) = \int_{-\infty}^{+\infty} f(t)\, e^{-j\omega t}\, dt$

Assume for the moment that $f(t)$ represents a voltage waveform; then looking at the types involved for each of these expressions, the right-hand side of this expression is:

$$= \int_{-\infty}^{+\infty} (Time \to Potential) \times \left(Time \to (Frequency \to Complex)\right) dt$$

which says that $f(t)$ is a function from Time to Potential and that $e^{-j\omega t}$ is a function from Time and Frequency to Complex (note that this has been grouped conveniently). Carrying the correct types through the multiplication and the applying the definite integral type above results in:

$$= \int_{-\infty}^{+\infty} \left(Time \rightarrow Potential \times \left(Frequency \rightarrow Complex \right) \right) dt$$

$$= \int_{-\infty}^{x} \left(Time \rightarrow \left(Frequency \rightarrow Potential \times Complex \right) \right) dt$$

$$= Time \times \left(Frequency \rightarrow Potential \times Complex \right)$$

$$= \left(Frequency \rightarrow Time \times Potential \times Complex \right)$$

which demonstrates, correctly, that the dimension for frequency-domain voltage (actually called a `Phasor_Potential` to be discussed in Section 3.4.2.1) is a complex-valued `Potential` per frequency (since `Time = 1 / Frequency`).

3.2.8.   Dimensions and Units

Another very important aspect which MHDL supports fully is the handling of dimensions and units, including definition of values with units, unit conformability though expressions and as function arguments, and allowance for user defined dimensions and units. The core MHDL language provides dimensions and units statements but, once again, all dimensions and units are defined in `Standard` for flexibility. For each dimension several unit statements can be made. The first of these defines the "base unit" for the dimension. Also note that each unit can have multiple variations in spelling. The following is excerpted from [LRM-94] in regard to physical length dimensions and units:

```
dimension length;
unit meter of length variations m, metre, meters, metres;
unit inch of length conversion 0.0254 meter variations inches;
unit foot of length conversion 0.3048 meter variations feet, ft;
unit yard of length conversion 0.9144 meter variations yards, yd;
unit mil of length conversion 0.00000254 meter variations mils;
unit micron of length conversion 1e-6 meter variations microns;
```

The first unit (required) declared for a dimension is called the *base unit*. The canonical form of expression values, as well as unit coercion in MHDL, uses the base unit preferentially so that `1 'foot' + 1 'm'` would be represented `1.3048 'm'`. Note that the arrangement in MHDL allows the addition of (user defined) units to existing dimensions as well as the addition of new dimensions with associated units. A highly abbreviated list of predefined dimensions and units is:

```
length          meter, inch, foot, yard, mil, micron
mass            gram, pound, ounce
time            second, minute, hour
current         ampere, amp, Amp, amps, Amps
temperature     Kelvin, centigrade, Rankene
```

```
force                    newton, dyne, pound_force
energy                   Joule, erg, calorie, electron_volt
power                    Watt,horsepower, dBm
Potential                Volt, ...
area                     square_meter, ...
inductance               Henry, ...
-- ...
```

Keep in mind that this is merely a highly abbreviated list for illustrative purposes. It is important to note that the first five of these are "base dimensions" while the last six shown are derived dimensions. For example, the code for inductance appears as:

```
dimension inductance = 'Potential * time / current'
unit Henry of inductance = 'Volt * second / ampere' variations H;
```

The MHDL base dimension and unit arrangement is directly based on ISO Standard 1000. The representation of Physical values makes use of a rational exponent for each base dimension. Indeed, it is the relationship of derived dimensions to the power set of base dimensions which allows dimensional/unit analysis. For example, values with dimension inductance would have a +1 in each of the Potential and time exponents and a -1 in the current category (and 0 elsewhere). As another example, area dimensions have a 2 in the length exponent. Rational numbers (i.e., one-third) rather than merely integers are allowed, since dimensions/units like 'dBm/Hz^(1//2)' arise. Rational numbers are handled exactly, both within units and in expressions; that is, 1//3 * 3 is 1 and not 0.999999 (note that // is a function used in infix form that takes two integers and creates a rational number). Note that attributes which represent physical quantities are "typed" using dimensions but given values of expressions with units. For example, an attribute might be declared to be of type time (e.g., attribute abc : time) and assigned a value of 10 msecs (abc = 10 'msecs';). An attribute declared to be of type Physical can represent any physically dimensioned quantity.

Finally, the ISO standard set of unit multipliers are also predefined via the MHDL multiplier statement:

```
multiplier yotta = 1e24;
-- ...
multiplier kilo = 1e3 variations K, k;
-- ...
multiplier micro = 1e-6 variations u;
-- ...
```

So that an expression like 10 'uH' * 14 'ka' evaluates to 140e-3 'Volt * second' in canonical, base-unit form.

3.2.9.   Other Features

Several remaining features necessary in analog modeling are next discussed in the five sub-subsections. Other than the description of interfacing to digital modeling via VHDL, this will conclude the description of general MHDL features.

3.2.9.1.   Tables and Generate

Many times information comes in the form of tables. Examples include published data
sheets, network parameters (e.g., S-, H-, Y-parameters), capture and (re-) use of
previous simulation results in future analysis, in the definition of distribution functions,
statistical (Monte-Carlo) results, etc. In recognition of this, MHDL provides a flexible
mechanism for use of table based data, namely the combination of a table data format
with a `generate` language construct to actually create values. The scope of a table is
within a structure, but for simplicity only the `table` and `generate` phrases are
shown in examples below. Multiple generate phrases can be used with a single table.
Consider:

```
table parms
title "Some attributes";
names attrib | value_typical |    min     |    max      ;
   {{ Vdc    | 5 'volts'     | 4.8 'volts' | 5.2 'volts' ;
      Zout   | 500 'ohms'    | 50 'ohms'   | 1 'kohms'   ;
      Idss   | 100 'ma'      | 99 'ma'     | 101 'ma'    ; }}
end parms;


for each row in table parms generate
    attrib = value_typical; -- just create typicals
end generate;


for each row in table parms generate
    attribute attrib : (Physical, Physical, Physical); -- type
    attrib = (value_typical, min, max);
end generate;
```

The following statements would be the equivalent effect of the first `generate`
statement: `Vdc = 5 'volts';` `Zout = 500 'ohms';` and `Idss = 100`
`'ma'`. In the second case, the `generate` creates statements including min and max
values. In table generates, the names of the columns are used in generate-body portion in
a general manner to access the table. While the table above was used to create a lists of
attribute definitions, the next fragments illustrate creation of an S-parameter matrix
function from a table:

```
table sparms_2x2
title "Measured by John Q. Public on 6/29/1994";
-- units of GHz for freq and degrees for angles assumed
names
freq       | s11m | s11a | s21m | s21a | s12m | s12a | s22m | s22a ;
{{ 1.0     | 0.98 | 32.  | 0.04 | 102  | 0.04 | 102  | 0.98 | 32 ;
   2.0     | 0.90 | 30.  | 0.06 | 104  | 0.06 | 104  | 0.90 | 30 ;
   3.0     | 0.85 | 28.  | 0.08 | 106  | 0.08 | 106  | 0.85 | 28 ;
   4.0     | 0.80 | 26.  | 0.10 | 108  | 0.10 | 108  | 0.80 | 26 ; }}
end sparms_2x2;


type Index_12 = Index (Int) -- "subtype" of Int
                where {Index(i) = i >= 1 && i <= 2 };
type Smatrix_2x2 = Gmatrix (Index_12, Index_12, Polar(Float));
               -- 2x2 matrix of polar floating elements
Smat :: Frequency -> Smatrix_2x2;
               -- a function from Frequency to an Smatrix_2x2
```

```
type Sentry_2x2 = (Physical,                       -- the frequency
                   Smatrix_2x2);                   -- the polar form data

ma = makearray( (Index (1), Index (2)) );
-- curried function for makearray which takes 2 args (2nd missing)

mp(a,b) = mkPolar(a, b 'degrees');

for each row in table sparms_2x2 generate
    Smat(freq 'GHz') = ma( [ma([mp(s11m,s11a), mp(s12m,s12a)]),
                            ma([mp(s21m,s21a), mp(s22m,s22a)])] );
end generate;
-- we how have an Smat function for frequencies defined in table!
```

The 2 by 2 matrix at each frequency is represented by an array of an array, which is a common technique for construction of multiple dimension data types from single dimension arrays (as in C). Note that although implementations can provide the generate feature in any way (provided it conforms to the semantic), MHDL is designed to allow compile time handling of table based generates. While "fixed" formats for, say, several typical cases of table usage could be provided within the language core, obviously the MHDL generate feature is general enough to enable handling of almost any case of data/information appearing in table form.

The function Smat defined above will return a 2 by 2 matrix for a supplied frequency argument for entries in the table sparms_2x2; it will however return an error for frequency arguments other than these. Note that this example omits the normalizing impedance/admittance which would also be needed. This highlights another requirement related to tabulated data, namely the need for a variety of *interpolations* (e.g. linear, spline). The technique used in this case is to alter the function definition to operate on indices and add an auxiliary data structure which maps these indices to frequency points. Thus, the actual *domain* of the function (that is the frequency points) becomes accessible as a value. This is necessary since, in order to interpolate, we need to know the next highest and lowest domain point for the function. The necessary changes are shown:

```
type Row_number = Int;
Freq_Points :: Row_number -> Frequency; -- func from index to f
for each row n in table sparms_2x2 generate
      -- Smat remains as above
      Freq_Points(n) = freq 'GHz'; -- new func "load"
end generate;

fr_lp ::Row_number ->                      -- Max size
        (Row_Number -> Frequency)  ->      -- func from index to f
        (Frequency -> Smatrix_2x2) ->      -- func from f to S
        Frequency ->                       -- interpolation f point
        Smatrix_2x2;                       -- resulting interpolation
```

```
fr_lp(table_len, freqs, matrices, fr) =
    if freqs(1) > fr || freqs(table_len) < fr
        then error("Frequency Extrapolation");
        else locate(1, table_len, freqs, matrices, fr)
            where {locate(first, last, freqs, matrices, fr) =
                if first+1 == last
                then interpolate(freqs(first),matrices(freqs(first)),
                                    freqs(last),matrices(freqs(last)),fr)
                else if freqs(middle) < fr
                    then locate(middle, last, freqs, matrices, fr)
                    else locate(first, middle, freqs, matrices, fr)
                        where {middle = div(first+last,2);};
            };

-- curried for the Smat function above!
Smat_lp = fr_lp(sparms_2x2.table_length, Freq_Points, Smat, );
-- Usage:
--    Smat_lp(1.5 'GHz') will produce [ [0.94,31 'degrees'], ... ]]
--    Smat_lp(5.0 'GHz') produces an error (interpolation only)!
```

The `fr_lp` function uses a standard (recursive) search to locate the next lower and higher points for the `interpolate` function (not shown). Similar techniques can be used to develop other interpolation functions. Note that since the table data ultimately (in the case above anyway) was used to define a function (with domain points as rows in the table), the technique demonstrated is applicable to any function, defined on a finite set of domain arguments; the only "added" information needed is the size of the set (in the case of the table this came from the standard attribute "`table_length`" on the table). Also, polymorphic versions (e.g., generalized on argument/result types) of interpolation functions are readily possible. Note that there is currently some discussion about encoding (and being able to access) the true domain of functions. Encoding and enabling access to the actual domain (of functions), would, for example, eliminate the need for the `Freq_Points` function to obtain interpolation information.

3.2.9.2. Multiple Domain Coupled Models and Association

As analog designers know, electrical performance is dependent on factors other than electrical domain characteristics. For example, a stable oscillator design may need to be compensated for temperature or packaging effects which derive from the physical design space (e.g., metal cover height, coupling of I/O pin wire bonds, ambient/self heating) may need to be represented. The initial step might be to provide multiple structures within each model as stated earlier. In this simplest case, separate electrical and thermal structures would be contained in each model. When the entire design hierarchy was created, the electrical and thermal structures would be completely separate and, of course, independent.

The next improvement on this would be to represent coupling between these structures, perhaps by writing expressions dependent on values from the alternate domain. While in some limited circumstances this might be enough, in general it is not since a one-to-one correspondence between electrical and thermal components and interconnects does not exist through all hierarchical levels. Therefore MHDL provides a more generalized means for "coupling" attributes through hierarchical structures resulting in a mechanism for loosely or tightly coupled models. This mechanism is called association.

The MHDL `associate` statement allows attributes of external structures, connectors, etc. to be used (assigned to or accessed as variables) within the local structure. The complexity of the interaction is entirely dependent on how the associated attributes are used, which might range from a "loosely" coupled description where a static thermal simulation result determines nominal temperatures for each electrical device to "tightly" coupled equations between electrical and thermal regimes. Further details will not be given here but reference [Saleh-94] presents an MHDL example of a self heating diode circuit example with coupled electrical and thermal models with differing structures.

3.2.9.3.   Specifications and Constraints

One of the strong user desires in [MRD-91] was for MHDL to support CAE beyond simulation, from the requirements phase through manufacturing and product support. On the synthesis side then, support for specifications is provided via the MHDL `constraint` statement. However, as the topic of this paper is modeling, only the use of constraints in modeling is discussed. While `constraint` statements can be used as the basis for specification or optimization goals, they are also used in models to express limitations or assumptions in models. Consider:

```
model mstrl  -- microstrip transmission line model
structure electrical
  connectors p1, p2, grd;
  attribute w,            -- line width
  l : Length;             -- line length

  constraint w_positive:(w > 0 'm') report "Width must be > 0";
  constraint l_positive:(l > 0 'm') report "Length must be > 0";

  -- ... rest of model ...
end electrical;
end mstrl;
```

The constraint has a label (e.g., `w_positive`) and report string (e.g., `Width must be > 0`) which is used to report violations to the user (along with the hierarchical name). The current semantic is for evaluation to stop as soon as any constraint is violated. Note that non-static checks can be performed as well, some MHDL fragments are:

```
attribute q : (time -> charge); -- time varying incremental charge
-- check that q is positive at ALL times
constraint charge_positive: (q(t) > 0 'coulombs')
        report "model not designed for depleted charge region";
--
attribute vbias : (time -> Potential); -- time varying voltage
constraint fast_v:(vbias(1 'msec') >= 1 'V')
          report "bias voltage not turned on fast enough";
```

Obviously, much more complex examples are possible but these serve to illustrate the mechanism for our purposes.

3.2.9.4. Primitives

Primitive functions provide a mechanism for accessing non-MHDL (i.e., C or FORTRAN) functions. As was already shown earlier, a function can be defined as `primitive "foreign_name"`. Although the actual details for argument passing and return value handling for the non-MHDL routine are left to the particular implementation, all implementations are obliged to provide this capability. Not only can rather low-level functions like `sin`, `cos`, `sqrt`, etc. be utilized via primitive functions, but complex analysis tools, or high-level functions (like FFTs or matrix operations) can be utilized. MHDL primitive functions accept any (possibly multiple) MHDL value(s), including functionals, as arguments. Although the language currently permits functionals as result return values for primitive functions, there are several implementation concerns that may prohibit this in the future.

3.2.9.5. I/O

Although primitive functions are the implementation basis in MHDL for input and output (I/O) of fundamental (e.g., `Int`, `Float`, `String`) types and binary coded files, input and output (I/O) is formalized in the classes `Text` and `Binary`. Instances for all of MHDL's base types of the "read string" (`ReadS`) and "show string" (`ShowS`) functions are collected in the class `Text`. An equivalent binary version is formed in the class `Binary`. Of course, user defined types can be instanced into the `Text` or `Binary` classes if custom I/O routines are desired, e.g.:

```
type user_type ...
instance Text(user_type) {
  -- define functions in class Text for user_type ...
```

Such features are useful in making use of data which is resident in files or databases and for writing evaluation results in particular form.

3.2.10. MHDL and VHDL

A very important requirement as defined in [MRD-91] is provision to describe mixed analog and digital systems. A variety of requirements led to forming MHDL based on a functional style and to a language separate from the digital description language VHDL [VHDL-93]. The means, then, for providing mixed system support is via a well defined interface between MHDL and VHDL. MHDL allows VHDL models to be directly instanced as components and also defines the meaning for instancing of an MHDL model as a VHDL `architecture` description. VHDL uses a language element called an `entity` which is used to define model arguments (called generics) and electrical connection points (called ports) and a language element called `architecture` to provide the actual model. Thus a complete model in VHDL requires an `entity-architecture pair`. Several architectures can be provided for an entity (each can be a different level of abstraction, for example) and the VHDL configuration mechanism is used to determine the entity-architecture pair in use. Although mixed-signal description/analysis is of prime importance, the main purpose of this paper is analog modeling, therefore the reader is referred to [Barton-94] and [Saleh-94] for details in this area and also to the discussion in Section 3.4.6.

3.3.   MODEL AND SIMULATION TYPES

Having defined MHDL language concepts relevant to modeling, next is a general discussion of analog modeling and simulation types and styles. This discussion is integrated with how MHDL supports these with language semantics and syntax. As was discussed in Section 3.1, a completely comprehensive model covering all aspects of an analog part is not possible or realistic. Thus models are developed with a particular purpose in mind; for our concerns here, this purpose is simulation.

There are many types of simulation of interest to analog designers, for example, transient analysis; small-signal (ac); interconnect/distributed element analysis; mixed-mode; d. c. operating point; harmonic balance; convolution-based mixed-mode solvers; physical device simulation; noise; electromagnetic simulation; etc. The examples of the next section really deal with the first four of these, although this is for space reasons only. Details on issues arising in these application areas will be discussed within the example descriptions. Before delving into the examples, some general discussion on the state of analog circuit/system solvers in relation to HDLs is in order.

3.3.1.   MHDL and Simulation Technology

Three major "modeling/usage scenarios" may be associated with an HDL in the course of its adoption, and these fit MHDL quite well. To make this discussion more definitive, consider the case of transient simulation via SPICE. In the first scenario, models are encoded outside of the language (via hooking SPICE models in via MHDL primitives) and MHDL serves mostly as a "netlist" conveying connectivity information but could provide some advanced features in argument definition, etc. In the second scenario, models are now written in MHDL. In both of these scenarios, however, equation extraction is limited and aimed at the modified nodal admittance (MNA) matrix solution within SPICE. In the third scenario, models are written in MHDL and a general-purpose higher-order integral-differential equation solver is used in conjunction with arbitrarily extracted equation sets. Of course, mixtures of these scenarios are also possible.

The first of these scenarios is obviously of little scientific interest, but is a plausible approach to making rapid use of the broad base of analog models available. The third scenario actually requires very little effort on the part of user or even extraction mechanism as everything is handled via a very generalized equation solver. The second case is the one in focus here, models are written in MHDL and we make use of "existing" solution (simulation) technology. The third scenario, of course, provides a long-term growth path for MHDL and recent developments [Fitzpatrick-95] indicate that this may not be that far off.

For transient solutions, typical non-distributed circuits result in linear, first-order integral-differential/algebraic equations. For example, in an admittance based formulation, linear inductors cause forms as $i = L \int V \, dt$ and capacitors cause forms like $i = C dV / dt$. Most simulators, including SPICE, are only designed for such first-order integral-differential equations (in comparison to higher-order ones where, say, derivatives squared/multiplied appear) and in fact *modify* the admittance formulation by mixing "state-variables" (currents and voltages in this case) so that inductor equations become $V = (1 / L) \, di / dt$ using Modified Nodal Analysis (NMA). Thus, only first-order *differential* equations are actually solved. So what are the ramifications of limited solvers at the HDL level?

As can be seen, MHDL provides a very generalized means (e.g., user supplied equations, attributes attached to various parts of the design) for describing terminal relationships or hardware characteristics. Rather than impose either syntactic or semantic restrictions as is done in some analog simulator input languages, there is no (MHDL) limitation regarding the description of, say, second order differential equations. Indeed, this is felt to be an important element of MHDL's language design, that is, focus on *describing* the hardware no matter what that may entail rather than constrain descriptions to what is "simulate-able" today. Also keep in mind that MHDL was designed with more than just simulation tools in mind but also analog synthesis, test, etc. As will be seen in Section 3.5, it is not difficult for tools (an equation solver) to determine if extracted equations conform to that simulate-able by them. Moreover, since MHDL assignments are really not procedural but are really *value definitions*, a good deal of rearranging of equations is possible by the tool on extracted equations. Thus, descriptions which "fit" within say a SPICE solution are simulate-able now, while those which do not are still describable and many are "transformable" to simulate-able form.

A final note here, integral-differential equations also require *initial conditions* to be complete. For example, the initial voltage of a capacitor may not be zero; even if it is zero, a proper description should state this. The current handling of this is provided by stating the initial condition as an added equation. For example:

```
EQ_CAP       : p1.i(t) == c*derivative(t)(p1.v(t) - p2.v(t));
EQ_CAP_INIT : (p1.v(0 'seconds') - p2.v(0 'seconds')) == 0 'volts';
```

provides the current/voltage relationship for a fixed, linear, time-independent capacitor with the initial condition of zero volts. While the first statement provides an equation that must be valid for all times (since ".i" and ".v" attributes are (Time -> Current) and (Time -> Potential) respectively), the second provides an equation valid only for t=0. Another point to note is that this example shows how to set initial conditions from *inside* a model. Such an initialization would be invoked for all uses (instances) of the model. More often, initial conditions would be defined (independently) for each model instance, for example as the result of a d.c. solution. In this case, the initial conditions would be defined using *hierarchical names* from a "root" model. That is, initial condition equations or value assignments would be written in the root description using the hierarchical names of the parts contained within (also see Section 3.5.1 concerning d.c. operating point).

Differential equations over a single independent variable are called *ordinary differential equations*, while MHDL provides for *partial differential equations*, that is the derivative function is not constrained to operate with only a single independent variable (which would normally be t). In such cases, complex *boundary conditions* are needed to specify the behavior along with (more complex) initial conditions. For example, resistivity over a surface may require description. While MHDL is indeed designed to support such cases, this is beyond the scope of this paper.

3.4.   ANALOG MODELING TECHNOLOGY WITH MHDL EXAMPLES

Next, examples are presented which illustrate MHDL in many analog and microwave regimes. These examples include a simple resistor-inductor-capacitor (RLC) model; transmission line modeling; a phenomenological model for a MESFET transistor; several circuits/systems with mixtures of various analog modeling types (mixed-mode); high-level signal flow modeling; and finally a system mixing analog and digital (mixed-signal). Following these examples, Section 3.5 briefly and generally discusses how MHDL interacts with CAE/CAD tools, making use of MHDL examples from this section.

3.4.1.   Simple Circuit Level

A fundamental circuit to describe is one containing a resistor, inductor and capacitor. This example provides a simplified illustration of how the functions described in Section 3.2.2 are used.

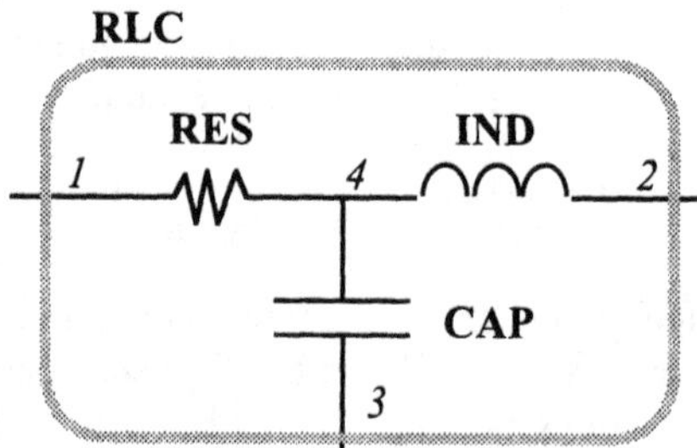

Figure 1: Schematic of a simple RLC circuit

Following are four MHDL models, one for each of the resistor, capacitor and inductor and a circuit called rlc which combines each of these models according to Figure 1.

```
model res
structure electrical
  connectors p1, p2;
  attribute R : Resistance;
  EQ_1 : p1.i(t) == (p1.v(t) - p2.v(t)) / R;
  p2.i(t) = -p1.i(t);
end; end;

model cap
structure electrical
  connectors p1, p2;
  attribute C : Capacitance;
  EQ_1 : p1.i(t) == C * derivative(t) (p1.v(t) - p2.v(t));
  p2.i(t) = -p1.i(t);
end; end;

model ind
structure electrical
  connectors p1, p2;
  attribute L : Inductance;
  EQ_1 : p1.i(t) == 1/L * integral(t) (p1.v(t) - p2.v(t));
  p2.i(t) = -p1.i(t);
end; end;

model rlc
attribute R : Resistance; attribute C : Capacitance; attribute L :
Inductance;
structure electrical
  connectors `1`, `2`, `3`, `4`;-- `4` is "internal node"
  components
    R1 :: res.electrical definitions R = 10 'ohms'; end R1;
    L1 :: ind.electrical definitions L = 10 'nH'; end L1;
    C1 :: cap.electrical definitions C = cv; end C1;
  end components;

  attribute cv : Capacitance;
  default cv = 1 'pF';
  `4`.ctype = electric;         -- node `4` is an electrical node

  connect R1.p1, `1`;
  connect L1.p1, `2`;
  connect C1.p1, `3`;
  connect R1.p2, L1.p2, C1.p2, `4`;
end; end;
```

In order to provide user-defined connection semantics, the applicable `Standard` code relies on two conditions. First that connectors have an attribute called `ctype` attached to them and that structures follow a naming convention. In the case of, say, KCL, connectors must be typed as `electric` and structures must be called `electrical`. Of course, these semantics are those provided by default in `Standard`, but the user is free to supply their own. In any case, the relevant `Standard` code for KCL/KVL is:

```
for each connector C generate          -- a "structure" generate
   if C.ctype == electric generate     -- "conditional" generate
      C_KCL : sum_chistr_objects(electrical, i) -
              sum_parstr_objects(electrical, i) == 0 'amps';
      C_KVL : C.v == children_connected_obj(C,v);
   end generate;
end generate;
```

Recalling Section 3.2.2, when this code is "evaluated" for `rlc`, the second of these 'sum' functions, `sum_parstr_objects`, would return nothing since there are no attributes attached to connectors at the `rlc` level. The `sum_chistr_objects` function returns the sum of ".i" attributes on connectors in structure `electrical` (note that `electrical` here is a name, which due to scoping rules represents the structure that it is currently in). Thus the following KCL equation gets generated:

```
rlc.C_KCL : res.p2.i(t) + cap.p2.i(t) + ind.p2.i(t) == 0 'amps';
```

Secondly, node `'4'` voltage is equated to all of the (children) connected connectors with an attribute `v` via the KVL rule using a function `children_connected_obj` (definition not shown). Ultimately then, the following set of equations are represented:

```
rlc.C_KCL : res.p2.i(t) + cap.p2.i(t) + ind.p2.i(t) == 0 'amps';
rlc.C_KVL : rlc.'4'.v(t) == res.p2.v(t) == cap.p2.v(t)== ind.p2.v(t);
```

As is shown, `sum_chistr_objects` returns the *hierarchical names* of connectors. Also note that in reality the equations generated would all be labelled as `C_KVL`, the `"rlc."` prefix was merely affixed for clarity. Note that although the 'extracted' equations are shown using MHDL syntax, these really only 'exist' within evaluation or CAD tools. Use of an MHDL syntax, however, makes it simpler to discuss.

3.4.2.   Distributed Components

As analog and digital circuit speeds increase, the need for accurate modeling of distributed elements becomes essential. This includes coupled and single transmission lines of many forms (i.e., IC or PCB printed lines), packaging elements such as bond wires and feed-throughs, etc. Indeed, modeling and simulation concerns which have been previously associated with the microwave domain will have increased prominence in even digital circuit modeling. For example, ringing and reflections on interconnects of a digital IC operating at > 200 MHz clock speeds requires accurate modeling of distributed elements (and major IC companies already have digital circuits with > 1 GHz clock speeds operating in their laboratories). Before delving into modeling of, and specifically MHDL modeling of, distributed elements, a brief review of transmission line theory is appropriate.

Distributed components like coupled and single transmission lines have always presented modeling and simulation challenges. The relationships governing the behavior of a generalized transmission line are described with the well known *telegrapher's equations*, which are:

$$\frac{\partial v}{\partial x} = -\left(L\frac{\partial i}{\partial t} + Ri \right)$$

$$\frac{\partial i}{\partial x} = -\left(C\frac{\partial v}{\partial t} + Gv \right)$$

where v and i are the time dependent voltage and current at any point along the line (x direction); that is they are total functions of both time, t, and position, x (e.g. $v(x,t)$). R, L, C, and G are electrical characteristics of per-unit-length series resistance and inductance and parallel capacitance and conductance, respectively. While these partial differential equations govern the line voltage and currents at any point of the line, we are most often interested (for modeling) in terminal characteristics for a line of some actual length, say l. Thus the boundary conditions for v and i (as functions of time) at positions 0 and l represent the model we are interested in. Efficient inclusion of such models would ideally require conversion of these partial differential equations into ordinary differential equations given the boundary conditions. Unfortunately, except for *lossless* lines, that is those with $R=G=0$, this has proven to be an elusive goal. This is why very early versions of SPICE contained only a lossless transmission line model.

In order to provide *lossy* transmission line models in transient simulation two methods are very commonly used. By recognition that the line voltages/currents actually behave as forward and backward traveling waves, the *method of characteristics* provides a means for simulation of lossy "RLCG" lines but requires storage of (and computation with) previous currents and voltages at the ends of the line (at $x=0$ and $x=l$). In practice, solvers generally store such information anyway so that this does not become a tremendous burden, but care does need to be taken to ensure that a small enough Δt is taken to ensure accuracy. A second even more simplified approach creates a lumped circuit equivalent to the distributed line. That is, multiple sections of lumped RLCG components are created to mimic the line. This method has major drawbacks. In order to achieve accuracy, a great number of RLCG sections needs to be created which gets worse as the length of the line is increased (in relation to the smallest wavelength or, equivalently, to the highest frequency or fastest transition applied). This of course increases the number of circuit nodes needed in the simulation which greatly degrades efficiency and increases memory requirements.

While even the introduction of loss impairs the ability to simulate distributed elements, the situation is in reality actually even worse. In the telegrapher's equations shown, the R, L, C, and G parameters are fixed (floating point) values. For many transmission lines of interest, including microstrip forms (e.g., IC or PCB traces) and coaxial configurations, these parameters are actually frequency-dependent. Some examples of frequency-dependency include: *dispersion* referring to the fact that waves of different frequencies travel at different speeds; and frequency-dependent loss (directly related to the R and G terms). Such conditions give rise to *distortion* in applied time domain signals. Within the microwave regime of analog modeling, a great deal of effort has been expended to create accurate models of such lines which are valid to high frequencies (e.g. >10 GHz). From an analog or mixed-signal circuit standpoint this might at first appear to be unnecessary. However consider a "modest" digital clock signal with a 10 psec rise time and 2 nsec period (50% duty cycle). The fundamental frequency component of this signal is 0.5 GHz (e.g. 1/2 nsec); via simple Fourier transforms, the ratio of the (voltage) magnitudes at 9.5 GHz to that at 0.5 GHz for such a signal is approximately 0.052, which is -12.8 dB. Thus the 9.5 GHz component of such a signal only 12.8 dB smaller than the fundamental and, depending on the design's sensitivity, accurate modeling at such a frequency may well be important.

Of course, the discussion in this section is mostly related to the *simulation* aspects, rather than the *modeling* aspects, of distributed elements. However, as might be concluded, an important fact relevant to modeling emerges, which is that frequency-domain models are essential for a complete analog HDL. There is simply no generalized way of creating time-domain models for, say, even a lossless LC transmission line with frequency dependent L and C, that is $L(f)$ and $C(f)$. Furthermore, there are other models which are more easily represented in the frequency domain (vs. the time domain) many of which have been accurately characterized by microwave engineers, including for example microstrip tee and cross junctions which appear in virtually every IC and PCB.

Fortunately, given that frequency-domain models will be included, there are mechanisms (e.g. convolution, pole-zero modeling/AWE) for inclusion of frequency-domain models in transient simulation. For example, current versions of SPICE (e.g. version 3f3 - May 1993) do indeed utilize convolution to enable the incorporation of non-frequency dependent lossy transmission lines. So the job then of MHDL is twofold, first provide a means for modeling frequency domain components and secondly to define the meaning (connection semantic) for the interconnection of such (it is also necessary to define the meaning of "mixed" interconnections but this is discussed in Section 3.4.4).

3.4.2.1.   A Frequency-Domain MHDL Model

In order to introduce the issue from a time-domain perspective, the discussion above centered about the RLCG model of a transmission line. More often, but equivalently, such a line would be characterized by its characteristic impedance, *Z0*, and propagation constant, γ (both complex-valued). The conversion between these sets of parameters is through simple algebraic expressions which can be found in rudimentary texts. The *network parameters*, namely the S-, Y- and Z-parameters, are also very commonly used in analog and microwave modeling. Again, there exists matrix transforms between all of these so for the moment consider them equivalent. Recall that Section 3.2.9.1 illustrated how S-parameter data is created from a table.

Just as MHDL functions of `Time  -> Potential` and `Time  -> Current` were used as the basis for time domain modeling and interconnection along with standard attributes v and i of these types, the standard attributes V and I of type `Frequency -> Phasor_Potential` and `Frequency  -> Phasor_Current` are used for this purpose in the frequency domain. Consider then a two-port (three connector) admittance description of a transmission line (see Figure 2):

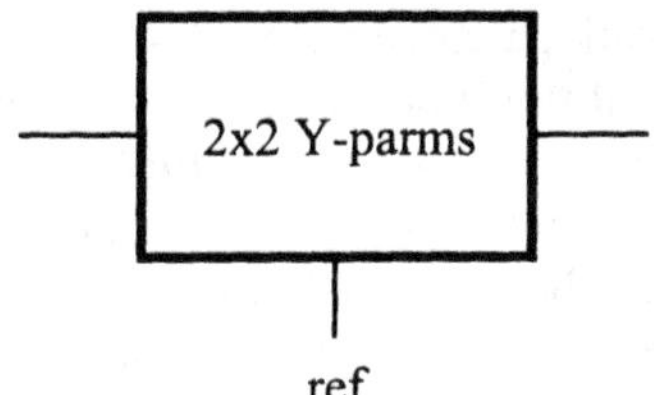

Figure 2: A 2x2 admittance block with reference point

```
model trl
attribute Y0 : Admittance;
attribute gamma : Complex;        -- model arguments
structure electrical
connectors p1, p2, ref;
attribute m_diag, m_offdiag : Admittance;
p1.ctype = fdomain;
p2.ctype = fdomain;
m_diag = Y0 * coth(gamma);
m_offdiag = Y0 * csch(gamma);
EQ1 : p1.I(f) == m_diag*(p1.V(f)-ref.V(f))
                 + m_offdiag*(p2.V(f)-ref.V(f));
EQ2 : p2.I(f) == m_offdiag*(p1.V(f)-ref.V(f))
                 + m_diag*(p2.V(f)-ref.V(f));
EQ3 : ref.I(f)== (m_diag+m_offdiag)*(ref.V(f)-p1.V(f))
                 + (m_diag+m_offdiag)*(ref.V(f)-p2.V(f));
end electrical;
end trl;
```

where Y0 is the characteristic admittance (1/Z0) and `coth` and `csch` are the hyperbolic cotangent and hyperbolic cosecant functions, respectively (in this case for `Complex` arguments). As is suggested by the names of the terms and the equation forms, these really conform to a matrix format.

The `Standard` supplied connection semantics for this example are quite simple. Since "I" and "V" attributes are merely phasor versions of "i" and "v", KCL/KVL also apply. Hence the phasor (frequency-domain) equations are extracted. As the `Standard` code for this is quite similar to the time-domain equivalent, it will not be shown. As may be noticed from the equations `EQ1` and `EQ2`, these are written in such a way that they will directly form the *nodal admittance matrix*. Of course, if all frequency domain models were written in this format, the extracted equations are readily usable in a small-signal or microwave simulator. As with time-domain equations, there is no restriction in MHDL to write equations in any particular manner (form). The user may introduce nonlinear terms, equations that (even rewritten) do not conform to admittance matrix form, etc. While this may adversely impact the ability to simulate the description *today*, such descriptions stand as challenges to future simulators. In the mean time, models can and will be written in such a way that allows maximum utility of existing simulation technology.

Linear relationships (given in models) between the standard phasor attributes I and V are, of course, admittances or impedances. As was mentioned earlier in this section, we also need to support the network parameters, (S-, Y-, Z-parameters). Y-parameters (Z-parameters could be equivalent used) are readily convertible to I and V relationships (nodal admittance equations) simply by *unnormalizing* and *floating* (that is adding a reference point) of the Y-parameters. As was stated, through standard matrix manipulations, any of these can be converted to any other. Thus, it is a simple manner to convert, say, Z- or S-parameters to Y-parameters (except for degenerate cases which are beyond the scope of this paper). These in turn are readily convertible to I and V phasor equations; and thus, in this manner all of the network parameters are supported. An example would consume a good deal of space so it will not be shown here. The reason for focus on the I and V phasors is because these are the basis for mixed-mode interconnection which will be shown in Section 3.4.4.

3.4.3.   Device Model

As another example, consider the phenomenological FET model presented in [Statz-87]. This reference describes a three terminal FET model designed to be compatible with SPICE and will serve as a more elaborate (over RLC) circuit-level MHDL example. The variable and function names from the paper are used in the model, except in cases where such names would conflict with a special MHDL name. Also names of physical constants are changed to longer names which are more descriptive. Also note that no attempt has been made to "optimize" this model in any way (e.g., through common sub-expression elimination) so that it more clearly demonstrates the ease of developing such a model in MHDL. As the model text is already quite long, only the portion of the model directly contained in this reference is included. While this makes it easier to compare, it does omit elements which are of importance (e.g., contact/parasitic resistances/inductances) if the model were to be actually used (see Figure 3).

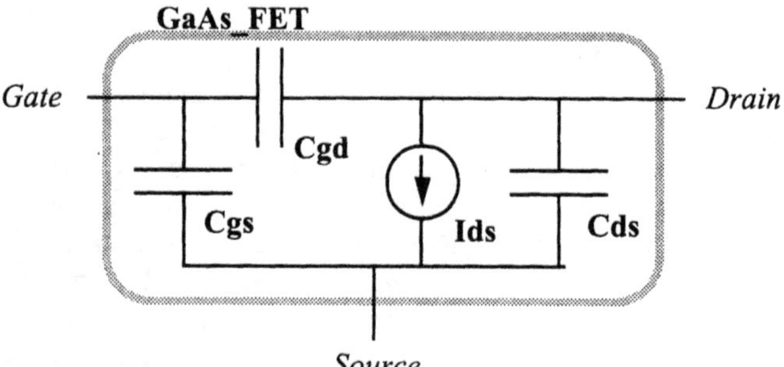

Figure 3:   Schematic of simplified GaAs FET model. While this model is insufficient for actual modeling purposes, it does utilize and demonstrate many MHDL features. Cds is a fixed capacitance, while Cgs and Cgd are time dependent.

From an MHDL modeling perspective, however, the missing elements do not require any MHDL techniques which are not already shown. As it stands, this example

illustrates handling of nonlinear, bias-dependent capacitances and drain-to-source
dependent current generator.

```
package GaAs_helper
-- ------------------------------------------- dimensions / units:
dimension Concentration = 'length^(-3)';
unit per_m3  of Concentration = 'meter^(-3)';-- base unit for dim
unit per_Cm3 of Concentration conversion 1e-6 per_m3;
-- ----------------------------------------------- attributes:
attribute k_GaAs : Float;
attribute q : Charge;

k_GaAs = 12.9;
q = 1.6021773349e-19 'Coulombs';
e0 = 1e-12 'farads/meter';
-- ----------------------------------------------- functions:
f1 :: Float -> Float -> Float;
f1(sr,d) = d^2 - sr^2 - 2//3*(d^3-sr^3);


end GaAs_helper;
-------------------------------------------------------------------------
model GaAs_FET
includes e_constants;
structure electrical
connectors gate, source, drain;
-- ------------------------------------------- model parameters:
attribute a,    -- channel thickness
          L,    -- gate length
          z     -- gate width (Cap Z in paper)
          : Length;
attribute Vt,   -- threshold voltage
          Vb    -- built-in voltage
          : Potential;
attribute beta,-- transistor beta
          b     -- model parameter (see ref.)
          : Float;
attribute N     -- donor density
          : Concentration;
-- ----------------------------------------- model parameter defaults:

default   Vb = 0.6 'volts';
default   N = 1e-17 'per_Cm3'; -- can put 'cm^(-3)' directly, too


-- ----------------------------------------- model internal values:


attribute Cgs, Cgd : (Time -> Capacitance);  -- time dep. caps
attribute Cds : Capacitance;                 -- fixed cap
attribute Vgs, Vds : (Time -> Potential);
attribute Ids : (Time -> Current);
attribute vel_saturated : (Time -> Boolean); -- device may enter/exit
                                             -- sat. region
attribute p : Float;
attribute W00 : Potential;
attribute Ws, sr,    -- "s" in paper conflicts w/ Laplace domain var
          d, p : (Time -> Float);
attribute Vsat : Physical;  -- generic physical (really potential,
                            -- see text)
Vsat  = 3.0e5 'volts/m' * L;-- potential at saturation
```

```
W00     = q*N*a^2/(2*e0*k_GaAs);
Ws(t)   = -Vgs(t) + Vb;
sr(t)   = sqrt( Ws(t) / W00 );
d(t)    = sqrt( (Vb + Vds(t))/W00 );
p       = sqrt( (Vb + Vsat)/W00 );

vel_saturated(t) = Vds(t) > Vsat;   -- carrier velocity saturated
                                    -- whenever ds voltage exceeds Vsat
Cgd(t) = if ( vel_saturated(t) )    -- if its now velocity saturated
    then
        k_GaAs*z*2*L/a*((2//3*(d^3-sr^3)-(d^4-sr^4)/2)*(1-sr) -
                    f1(sr,d)*(sr-sr^2) ) / f1(sr,d)^2
    else
        k_GaAs*z*(f3(t) + f4(t))
            where {
                zeta = Vsat*L/W00;
                f1sp(t) = f1(sr(t), p);
                f2sp(t) = f2(sr(t), p);
                f1spP(t) = 2*p*(1-p)*pP;
                f2spP(t) = p(t)*f1spP(t);
                pP = Not_Shown;
                L1 = From_Another_Paper;
                L1P = Not_Shown;
                f3(t) = -2*W00/f1sp(t)^2*(f1sp(t)*f2sp(t)*L1/a +
                        f1sp(t)*f2sp(t)*L1p(t)/a -
                        f1spP(t)*f2sp(t)*L1/a);
                f4(t) = -2*W00*(L2*pP(t)/a - L1P*(p/a+
                        zeta*sinh(pi*L2/(2*a))/(2*L)));
                };

Cgs(t) = Not_Shown(t) 'pF';-- much the same as Cgd
Cds     = From_Another_Paper 'pF';
Ids(t) = beta*(Vgs(t) - Vt)^2 / (1 'volt' + b*(Vgs(t)-Vt));

components
    C_gs :: tcap.electrical definitions c = Cgs; end;
    C_ds :: cap.electrical definitions c = Cds; end;
    C_gd :: tcap.electrical definitions c = Cgd; end;
    Igen:: cs.electrical definitions g = Ids; end;
end components;

-- connect Vgs/Vds to ports (currents handled via connects)
EQ1 : Vgs(t) == Gate.v(t) - Source.v(t);
EQ2 : Vds(t) == Drain.v(t) - Source.v(t);

connect C_gs.p1, C_gd.p1, Gate;
connect C_ds.p1, Igen.Source_Top, C_gd.p2, Drain;
connect C_ds.p2, Igen.Source_Bot, Source;

end electrical;
end GaAs_FET;
```

As was intended, this example demonstrates several interesting aspects of MHDL.
However, note that the terms Not_Shown or From_Another_Paper are used where
provision of the actual expressions would not provide any additional challenges but
merely fill space. First, notice the clear handling of functional types, the MHDL
(functional) type system provides direct (and type-checked) handling (creation/use) of

time domain functions. The function `d(t)` was defined explicitly in terms of the argument `t`. However, consider the assignment `d = sqrt((Vb + Vds)/W00);`. In this case, the term `Vds` is known to be of type (`Time -> Potential`) so that ultimately the coercions that take place result in a type (`Time -> Float`) for the right hand side. For convenience, the special names (in this case `t` of type `Time`) are treated specially in simple expressions so that the correct functional form is obtained; thus,`d = sqrt(Vb + Vds)/W00);` and`d(t) = sqrt((Vb+Vds(t))/W00) ;` are the same.

Secondly, note the use of type inference (mentioned in Section 3.2.6) for functions defined, for example, in the `where` clause of `f1sp(t)` -- this saves a considerable amount of extra input. Third, notice the dimension/unit handling in the expression for `W00`, `q*N*a^2/(2*e0*k_GaAs)`. Ignoring constants and multiplier conversions, the dimensions of this expression is `charge*length^(-3)*length^(2) / (farad*length^(-1))`. MHDL would convert each of these into the power set of "base dimensions". Since all of the `length` dimensions will factor out anyway, just consider `charge / farad`. MHDL would convert these into the base dimensions, and discover a match with `Potential` (which is the correct dimension for W00).

A fourth point to consider is the declaration for `Vsat` which is `Physical`. While the dimension for `Vsat` is actually a `Potential`, the model author chose to use the "generic" type for physical quantities. The type `Physical` can represent any physical value and can be used to avoid mandatory creation of dimensions/units, in this case for `Vsat`.

A fifth point related to the example is consideration that this model was designed to be "simulate-able" and hence translates quite simply. Since MHDL does not "execute" statements, note the use of the `"if"` expression in defining `C_gd`. The expression for `Cgd(t)` when the device is saturated actually derives from a partial derivative (of total gate charge, Q_g, with respect to gate-to-drain voltage, V_{gd}). Fortunately, this was able to be done explicitly, but suppose this was not the case. In MHDL such a non-time derivative could be represented directly (i.e., `derivative(Vgd)(Qg(Vgd))`). Partial derivatives would be represented in a similar manner; for example `derivative(x)(func(,,,x,,,))` represents the partial derivative of `func` with respect to its fourth argument. Moreover, *implicit equations* (where a value or set of values are determined via solution of one or more equations) is directly possible in MHDL. Simulators generally code such cases directly into models (say use a Newton-Raphson solver to determine the independent variables value from an un-invertible function) and MHDL provides a means for direct and precise representation of such cases.

Notice the use of the `tcap` models for `C_gd` and `C_gs`. Unlike `C_ds` which is a fixed capacitance, these are time-varying capacitances. Via MHDL's type checking, it would be a type error to pass, say, `C_gd` to the cap model as it expects a "C" value of `Capacitance`, not (`Time -> Capacitance`). Therefore the `GaAs_FET` model uses the `tcap` model which provides for a time dependent capacitance; its definition is:

```
model tcap
structure electrical
   attribute c : (Time -> capacitance);
   connectors p1, p2;
   EQ_1 : p1.i(t) == derivative(t) (c(t)*(p1.v(t) - p2.v(t)));
   p2.i(t) = -p1.i(t);
end; end;
```

Other than the differing declaration for c, the only other difference is that c has moved inside the derivative. Of course since it is no longer constant, it cannot be moved outside of the derivative function. Of course, as the tcap model is a more generalized version of cap, the cap model isn't really necessary. That is, a "constant" time function can be passed to tcap to achieve a fixed capacitor. Such a function would look like \ (t) -> 10 'pF'; (note that the "\" is used to represent a function which has no name) which is the (non-strict) function which always returns 10 'pF'. While this is possible, providing cap long with tcap makes for better ease of use. As might be noticed, the functional basis, along with dimensional analysis, of MHDL allows handling of this in a simple and natural way.

3.4.3.1.   Equation Extraction

While Section 3.4.1 showed KCL/KVL for an internal node, here the GaAs_FET model's external connections are shown.

```
Gate.i(t)      == derivative(t)(C_gs(t)*(Gate.v(t) - Source.v(t))) +
                  derivative(t)(C_gd(t)*(Gate.v(t) - Drain.v(t)))
Drain.i(t)     == derivative(t)(C_gd(t)*(Drain.v(t) - Gate.v(t))) +
                  C_ds*derivative(t)(Drain.v(t) - Source.v(t)) - I_ds(t)
Source.i(t)    == derivative(t)(C_gs(t)*(Gate.v(t) - Source.v(t))) +
                  C_ds*derivative(t)(Source.v(t) - Drain.v(t)) + I_ds(t)
where {
      C_ds     = "from model";
      C_gs(t)  = "from model";
      C_gd(t)  = "from model";
      V_ds(t)  = "from model";
      I_ds(t)  = "from model";
      V_gs(t)  == Gate.v(t) - Source.v(t);
      V_ds(t)  == Drain.v(t) - Source.v(t);
      }
```

In summary then, a "significant" analog model involving time dependent components and characteristics (dynamically occurring velocity saturation) has been shown. As can be seen, these resultant equations are implicit in the terminal relationships, for example the V_gs function is known from the model expressions and this is actually the gate-to-source voltage difference. Thus the terminal voltage/current relationships are only "known" implicitly. If necessary, the model could be readily re-written explicitly in terms of its terminal voltages (see Section 3.5 for additional discussion).

3.4.4.   A Mixed-Mode Case

Of prime concern to MHDL is the ability to represent (and utilize) *mixed-mode* connections, which is the interconnection of models represented in different ways or domains. While the issue of such interconnections actually concerns the whole gamut of such connections, for space reasons only a mixed frequency and time domain example is discussed here (note the mixed analog/digital connection is described below). The example of Section 3.4.2.1 depicted a frequency-domain model and discussed how any of the major frequency-domain descriptive types (e.g. S-, Z-,Y-parameters) could be converted to nodal admittance forms (based on `I` and `V` attributes which are `Frequency` to `Phasor_Current` and `Phasor_Potential` functions, respectively) and also how such a nodal admittance matrix could be extracted.

As was shown by determination of the result type for a Fourier integral in Section 3.2.7, the correct dimensions for `Phasor_Potential` and `Phasor_Current` for the Section 3.4.2.1 example are `'volts / hertz'` and `'amps / hertz'`, respectively. Connectors with these attributes are `fdomain` types, while connectors with time domain voltages and currents are typed `electrical`. In any case, the mixed interconnection of such is shown in Figure 4.

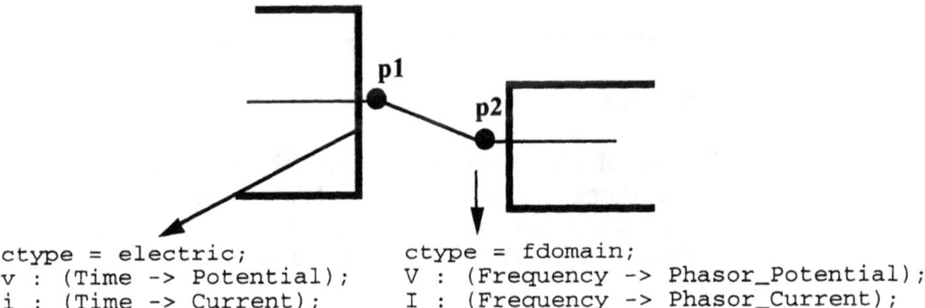


```
ctype = electric;                    ctype = fdomain;
v : (Time -> Potential);             V : (Frequency -> Phasor_Potential);
i : (Time -> Current);               I : (Frequency -> Phasor_Current);
```

Figure 4:   Connectors of two different types are interconnected. Connector p1 has a type `electric` and attributes v and i defined for it, while connector p2 has a type `fdomain` and attributes V and I defined for it.

It is important to note that *connectors* are typed -- not models. That is a single model/structure can contain connectors of varied type; otherwise it would not be possible, for example, to write user-defined models aimed at explicitly bridging the various descriptive domains. The connection semantics of MHDL however, allow, in many cases, the direct connection mixed connection types. Remembering that the role of an HDL is to *describe* an object, not to *perform* operations on it (which is the role of CAE tools), strictly speaking, MHDL does not really need to "do" anything in regard to this interconnection other than to give it precise meaning.

The requirement is met via the forward and inverse Fourier transforms. Although, for example, the forward/inverse Fourier `integrals` could be written in MHDL (refer to Section 3.2.7), they are actually provided as `primitive` functions [LRM-94]. The connection semantic is precisely defined through these transforms which relate v to V and i to I attributes. Having precisely defined the mathematics of such a mixed interconnect, the second issue now is the simulation ramifications of such a mixed connection. This is treated further in Section 3.5.2.

3.4.5. Signal-Flow Level

Moving towards a much higher level of abstraction, this section will develop a signal-flow model. The "signal" in this case is a time dependent series of floating point values. In this simple case, models are represented as transfer functions and allowances are made for impedances mismatches. These considerations are still at a fairly abstract level, that is the mismatch merely degrades the signal amplitude. But this is a reasonable demonstration for a high level model. The MHDL text below defines three models and a package which is used within each of these models as shown in Figure 5.

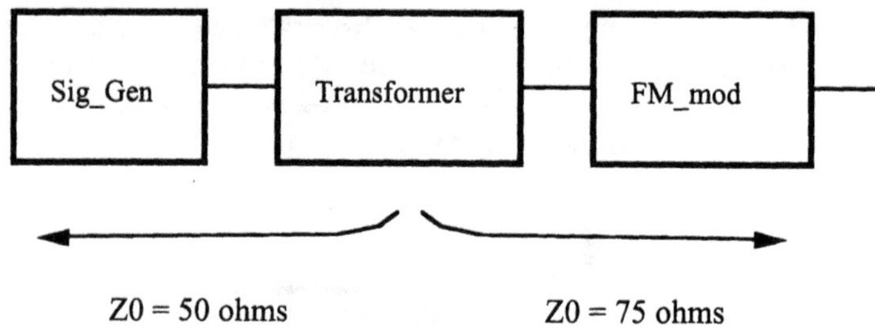

Figure 5: Block diagram of simple system-level model consisting of a signal generator, transformer and simple FM modulator. The signal generator (Sig_Gen) has a 50 ohm characteristic output impedance while the FM_modulator (FM_mod) has 75 ohm characteristic input and output impedances.

```
package my_signal
type My_Sig = (Time -> Float);      -- type for my signal
structure any
  connectors inp, out;
    attribute my_sig : My_Sig;         -- std attribute for my signal
    attribute Z0in, Z0out : Impedance;

    default inp.Z0in = 75 'ohms';      -- default connector impedances
    default out.Z0out = 75 'ohms';

    mismatch = if (z1 > z2) then sqrt(z2/z1)
                            else sqrt(z1/z2) -- always <= 1
                  where { z1 = inp.Z0in;
                          z2 = inp.Z0out; };
end any; end my_signal;
-- -----------------------------------------------------------------------
model Sig_Gen
includes my_signal;
structure any
  connectors out;
  attribute the_signal : My_Sig;
```

```
    out.ctype = sigflow;    out.dir = outdir;
    default the_signal(t) = sin( 10 'KHz' * t );
    out.my_sig(t) = the_signal(t);
end any; end Sig_Gen;
-- ----------------------------------------------------------------
model Transformer
includes my_signal;
structure any
   connectors inp, out;
   inp.ctype = sigflow;    inp.dir = indir;
   out.ctype = sigflow;    out.dir = outdir;
   out.my_sig(t) = mismatch * inp.my_sig(t);
end any; end Transformer;
-- ----------------------------------------------------------------
model FM_mod
includes my_signal;
structure any
   connectors inp, out;
                                     -- MODEL PARAMETERS
   attribute cf,                     -- center freq
            freq_slope : Frequency;  -- freq sweep per input sig
   attribute scale      : Float;     -- scale of output

   inp.ctype = sigflow;    inp.dir = indir;
   out.ctype = sigflow;    out.dir = outdir;
   out.my_sig(t) = mismatch*scale*sin((cf+freq_slope*inp.my_sig(t))*t);
end any; end FM_mod;
-- ----------------------------------------------------------------
model Root
includes my_signal;-- include base package
structure any
   connectors p1, p2, p3;
   components
     Part1 :: Sig_Gen.any definitions
                    attribute the_signal : My_Sig;
                    the_signal = \(t) -> cos( 5 'KHz' * t );
                    out.Z0out = 50 'ohms';
                    end Part1;
     Part2 :: Transformer.any definitions
                    inp.Z0in = 50 'ohms';
                    end Part2;
     Part3 :: FM_mod.any definitions
                    attribute cf, freq_slope : Frequency;
                    attribute scale : Float;
                    scale = 2;
                    cf = 400 'MHz';
                    freq_slope = 15 'MHz'; -- MHz / "Float"
                    end Part3;
   end components;

   connect p1, Part1.out, Part2.inp;
   connect p2, Part2.out, Part3.inp;
   connect p3, Part3.out;
end any; end Root;
```

The package my_signal is used (included) in each model for several purposes. The first is to establish a type (My_Sig) for the signal (in this case just a simple time to floating point value) along with an attribute of this type (my_sig) to consistently attach to the connectors of the models. Secondly, the package `my_signal` defines a utility function called `mismatch` to be used in subsequent models. As can be seen, this function defines a floating point value in terms of two impedance values, `z1` and `z2`. A key element in the definitions for `mismatch`, `z1` and `z2` is the structural combination rule discussed in Section 3.2.1 which is that included structures of the same name are 'combined'. For example, this means that the connector `"inp"` used in the assignment for `mismatch` in the package/structure `my_signal.any` is one in the same with the connector `"inp"` in each (derived) model (e.g., `Sig_Gen`, `Transformer`, and `FM_mod`). Thus the definition of `inp.Z0in` is applies to the correct value in each model. In all cases, each model declares the type of their connectors to be `sigflow` and also declares the directionality of its connectors. Each of these models is then instanced in the model `Root` to form the total description. Another point of interest is the definition of the functional for `the_signal` in `Part1`, which actually overrides the default definition in the signal generator; also note that attributes must have a known type and hence the need to declare those not already known in each part definition.

3.4.6. Mixed-Signal (MHDL/VHDL)

An important aspect to an analog HDL is the interfacing/integration with a digital HDL, namely VHDL, in order to facilitate the description of mixed-signal circuits and systems. Since the goal of this paper is analog modeling, only a very brief example will be used to demonstrate MHDL - VHDL integration. An MHDL part can actually be an instance of a VHDL model (entity-architecture pair) and VHDL can instance an MHDL model-structure. In both cases, the MHDL language definition [LRM-94] determines the meaning of mixed connections and does not require any changes to the VHDL definition [VHDL-93]. A very simple example is one where a digital (VHDL) element controls an analog circuit (see Figure 6).

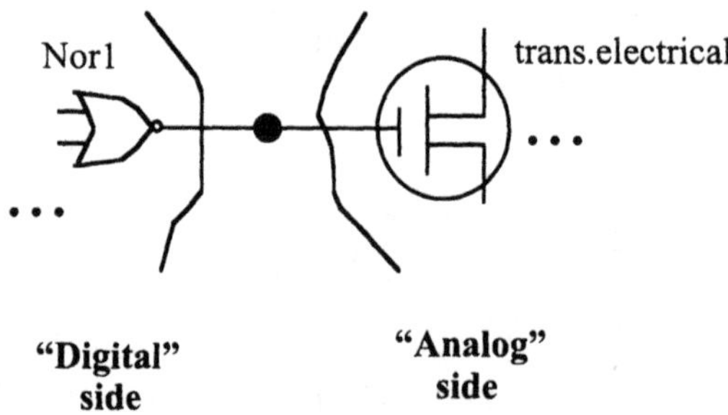

Figure 6: A very simple mixed-signal interconnection.

The MHDL/VHDL fragments are as follows:

```
------ in MHDL ------
model trans
structure electrical
  connector gate, source, drain;
  ...
```

```
model main
include VHDL_package;
structure mix
  connector mixed_connect;
  analog_part :: trans.electrical •••
  dig_part    :: nor1.example •••

  connect analog_part.gate, dig_part.c;
  •••

------- in VHDL ------
entity nor1 is
  port (a, b: in BIT; c: out BIT);
end nor1;
architecture example of nor1 is
begin
  c <= a nor b after 1ns;
end example;
```

which shows how (using the default connection semantics) the mixed-signal case is handled. Of course, this is a very simple case, MHDL freely allows the instancing of VHDL design entities as elements. There is no need for the user to separate the circuit into digital and analog parts (this was shown for clarity in the figure) as this is inherent to the description. Also all VHDL models can be used in the analog regime without modification, as the (electrical) information needed to fully define the interconnect comes from the MHDL "side" which utilizes threshold/ramp functions for level values (in the default case).

3.5. CAD/CAE TOOL INTERACTION

While the prime focus of MHDL related efforts have been in the definition of a language, consideration for how it will be used is also critical. The general concept for MHDL implementation is the provision of an "MHDL front-end compiler" coupled with a variety of CAE tools. Such a front-end compiler would eliminate the need for each CAE vendor to develop common elements needed to implement MHDL. The front-end compiler would provide several features, for example, basic parsing and compilation to an "object" form, support for libraries/separate compilation, handling of functional programming semantics and evaluation, as well as provide other general-purpose facilities. Two such tools are currently under development.

As has been shown, the basic technique is to provide engineering connection semantics in the form of run-time (and hence modifiable/extendable) code. Based on such connection "rules", the extraction of necessary equations, along with supporting assignments as defined in models (e.g., see Section 3.4.3.1), is a primary mechanism for simulation. As was pointed out in Section 3.3.1, time or frequency domain extracted equations are not necessarily in any particular form, but are, in general, higher-order arbitrary-degree integral-differential equations. For example, a second order differential equation could be reduced to a pair of coupled first order differential expressions or a change of state variables can transform an integral-differential equation to differential (only) form. Such transformations are only guaranteed to be semantically equivalent since MHDL functions and attributes are referentially transparent and pure. If they weren't, the number of executions and their order may have an effect and therefore different results would be possible.

In cases where equations are not needed a tool could make use of a front-end compiler/evaluator to essentially provide "simulation" directly. For example, the evaluation of the attribute `Part3.out.my_sig` in the `Root` model in Section 3.4.5 for a sequence of time point would constitute a "simulation". Alternatively, a CAD tool might extract the transfer functions defined in each model, apply the specified connection semantics, and perform the simulation (with the same result of course).

3.5.1.   Tool Results Return

Another requirement in the design of MHDL is the ability to document and possibly re-use results of previous simulations (as well as the results of other tool operations) [MRD-91]. As examples, the results of a d.c. solution need to be used as the starting point for a transient solution (assuming that both of these are done in separate steps); or a small-signal analysis may be captured as a admittance matrix; or the results of a transient solution should be "attachable" to the design; etc. Like any other hierarchical language, the MHDL source code is "unelaborated", that is, it is a series of description units which are "instanced" (perhaps multiple times) within each other. For example, the current flow through a capacitor model does not exist "once", but for each independent usage (instance) of the capacitor. It is therefore impossible to annotate the capacitor model, for example, with its solved current value(s) (except in the trivial case where only one capacitor appears in the entire design hierarchy). As was alluded to in Section 3.3.1, the "answer" lies in returning results to the "main" description using hierarchical names. For example the d.c. solution might be "returned" by a tool in the form:

```
main.struct ... part.cap.electrical.v(0 'sec') = 0.15 'mV'
```

A transient voltage might be returned as a `table` (along with appropriate `generate` statements) and attached to say an attribute `"v_solved"` which, like the standard attribute `"v"`, has a type (`Time -> Potential`). The table would contain the full hierarchical name and time points/values of the result. As another approach, a table could be used to form the signal generated by the `Sig_Gen block` in Section 3.4.5.

3.5.2.   Simulating the Mixed-Mode Case

Section 3.4.4 demonstrated how MHDL precisely defines the meaning of a mixed-mode interconnection (specifically the case of frequency and time domain). Since we are also interested in how MHDL is used, we also need to consider simulation ramifications for such descriptions. The first thing to is consider is what type of simulation is being performed. Two obvious cases are a small-signal (a.c.) simulation may be carried out (which entails "translation" of the time-domain model to frequency-domain), or a transient simulation might be desired (where now the frequency-domain model needs to be "translated" to the time domain) -- each of these is briefly discussed next. Keep in mind that the primary purpose here is to discuss language implications not the description of mixed-mode simulation technology.

3.5.2.1.   Small Signal (a.c.) Analysis

While the case of strictly fdomain connectors, including the extraction of simulate-able nodal admittance equations, was discussed in Section 3.4.2.1, here the issue is the "handling" time-domain models. Deriving a small-signal model from transient equations is a well-known technique, generally speaking time derivatives become $j\omega$ terms and time integrals become $1/j\omega$ terms. Such techniques rely, however, on very strong ties between built-in simulator models and the simulation procedure itself. The first scenario described below indeed aims to enable this level of operation, while the second provides more generalized capabilities.

In the first case, the simulator extracts the full set of transient equations (electric typed connectors) along with the frequency domain models (with fdomain connectors). The next step is to verify that, in fact, the (default) connection semantic is in place, namely KVL/KCL and that mixed interconnects are indeed subject to the default forward and inverse Fourier transforms. The Fourier transform in this case is not actually applied, rather a d.c. solution is performed. For fdomain models this would mandate that data be available (or assumed) for $f=0$. Given the d.c. point, the time-domain equations are linearized as discussed above to form frequency (equivalent to fdomain) domain equations which can be "added" (via KVL/KCL) to the fdomain models. Thus frequency domain nodal admittance equations are created and extracted for the complete description and a small-signal analysis is readily performed.

The method just described will (readily) work only in cases where the assembly of the time domain equations results in ordinary, first-degree (linear), first-order time derivative/integral expressions. In these cases, a good deal of efficiency is garnered due to the avoidance of explicit transforms (while the language semantic is still upheld, of course). While this case encompasses a large class of circuits and systems, obviously there is no guarantee that such equations will result as the user is able to write arbitrary models. Thus the second alternative is for the simulator to more "directly evaluate" the description. Within this operational regime, several possibilities exist, a complete interface between all mixed connections could be formed, or each model could be treated on an individual basis. The "treatment" here is the application of the (forward) Fourier transform to the terminal i/v relationships to form terminal I/V relationships. This could be done symbolically, but would more likely be done numerically except in idealized circumstances. Even the numerical transformation is, in general, quite difficult. The Fourier integral (shown in Section 3.2.7) requires integration over the entire domain (from negative to positive infinity, to t really) of the function being transformed. Thus unless the function has a known, finite domain over which it is non-zero, it is not generally possible to compute this integral. Causality arguments aid somewhat here, but there is no guarantee that the user has provided a casual model. Nevertheless, from a pragmatic point of view, simulators exist which do operate with such mixed mode cases, and MHDL provides the information content needed for their operation.

3.5.2.2.   Transient Analysis

Looking at the somewhat opposite case, consider the inclusion of `fdomain` models in a transient simulation, along with `electrical` models (with connectors of type `electric`). Section 3.4.1 showed the extraction of the appropriate time domain equations for such cases. Similar to the second case of the previous section, a "direct evaluate" method is possible where inverse Fourier transforms are used (along with convolution) which integrate frequency domain models into a transient simulation. If the `fdomain` model is *band-limited* then the inverse Fourier integral can theoretically be computed exactly, or at least computed to an arbitrary degree of accuracy. Alternatively, some approximate methods may be employed. For example, the simulator could fit Laplace domain (in terms of `s`) polynomial pole-zero functions to the terminal relationships actually supplied in the models, in a method used with the Asymtotic Waveform Evaluation (AWE) technique. Such a model is then directly incorporable in the transient solution. Keep in mind that this is an approximate method with an accuracy limited to the accuracy of fit (although in the case of AWE it can be shown that greater accuracy can be obtained using higher-order moments; another important aspect of AWE is that the approximation avoids introduction of non-causality/instabilities which are possible with other moment truncation methods).

In these mixed-mode cases, it is important to note that MHDL has indeed fulfilled its primary objective, namely the accurate description of the mathematics of the situation and enabled a range of possibilities for usage of this description. Remember that many other "mixed-mode" cases are possible but only the time/frequency domain case has been addressed in this section.

3.6.   CONCLUDING REMARKS

MHDL language features have been defined from an analog and mixed-signal modeling perspective. Since this is in fact a major purpose of MHDL, actually little of MHDL was not discussed here. Some of the features not discussed are lists; arrays and matrices; record types and field selection; geometric description; information hiding (which, for example, allows "privatization" of model internals); statistical values; and other simulation types (e.g. noise, harmonic balance, Laplace and Z domain). As can be witnessed, great care is necessary to ensure precise engineering meaning is given to what is really a computer language (i.e., proper typing of expressions, handling of dimensions/units, provision of expected features like signals in a convenient form, familiar terminology, etc.).

MHDL describes analog behavior without resorting to execution of procedural statements. Rather than modeling in terms of updating of state-variables defined within models, MHDL captures signals and behaviors directly as language (functional) types. By avoiding state and procedure within models, MHDL does not constrain the solution of the (analog) equations to a particular method. It is thus possible to support a wide range of simulator types (i.e., transient, small-signal, harmonic-balance). It allows the definition and integration of a set of "vertically" abstract models through the ability to alter default connection semantics. In addition to the `Standard` package, which defines all of the functions, types, etc. that must be supported by any MHDL tool, the `MHDL_Standard` library contains a set of models which provide a base set of models from which to work (note that this library is currently under development).

A final point to note is that MHDL is still under development (by the MHDL Study Group) within the IEEE standardization process, although language changes appear to have stabilized at this point. As an example of something that may change notice that the current connection semantics are based on typing of connectors. An alternative approach would base connection semantics based on which attributes are attached to connectors (i.e., if `I` and `V` are used then this must be an `fdomain` connector). While changes are still on-going, this paper, of course, documents the language as it stands today. I would like to thank Doug Dunlop of Intermetrics for useful suggestions as well as the for the complete code related to the interpolation of S-parameters contained in Section 3.2.9.1. I would also like to thank the entire MHDL Study Group for all of the intriguing and stimulating discussion generated in the course of defining MHDL.

REFERENCES

[Barton-94]   D. Barton, "Issues in Mixed Mode Simulation Using VHDL and MHDL", Proc. of Int. Conf. on Simulation and HDLs (SHDL'94), Tempe, AZ, January 1994, Paper 2-2, publ. by Society for Computer Simulation (SCS).

[Fitzpatrick-95] D. Fitzpatrick, K. Kundert, "Simplified Formulation Methodology for Describing Analog Behavior", Proc. of Int. Conf. on Electronic HDLs (ICEHDL'95), Las Vegas, NV, January 1995, Paper III.4, publ. by Society for Computer Simulation (SCS).

[LRM-94]   "MHDL Language Reference Manual (LRM)", Version 2.0, Draft Working Document of IEEE SCC-30, MHDL Study Group.

[MRD-91]   "MHDL Requirements Document", Edition 1, August 1991, publ. Army Reserach Laboratory, Fort Monmouth, NJ.

[Saleh-94]   R. Saleh, D. Rhodes, E. Christen, B. Antao, "Analog Hardware Description Languages", Proc. of Custom Integrated Circuits Conference (CICC'94), May 1-4 1994, paper 15.1.

[SRD-94]   "SCC-30 Requirements Document", Version Draft 1, Draft Working Document of IEEE SCC-30.

[Statz-87]   H. Statz, P. Newman, I. Smith, R. Pucel, H. Haus, "GaAs FET Device and Circuit Simulation in SPICE", IEEE Trans. on Electron Devices, Vol ED-34, No. 2, Feb. 1987, pp. 160-169.

[TIRM-94]   "MHDL Tool Integrator's Reference Manual", Version 2.0, March 1994, publ. Intermetrics Incorporated, McLean, VA.

[VHDL-93]   "VHDL Language Reference Manual/1993", IEEE Standards.

[Zhou-91]   W. Y. Zhou, H. W. Carter, 1991, "AnaVHDL: A Mixed Mode Simulation Capability Using SPICE and VHDL", Proc. of VHDL User's Group Conference, Spring 1991.

4

MODELING AND SIMULATION OF ELECTRICAL AND THERMAL INTERACTION

H. A. Mantooth, E. Christen

Analogy, Inc., P.O. Box 1669, Beaverton, Oregon 97075, USA

ABSTRACT

The key to a successful electro-thermal simulation is the fact that the system to be simulated is a closed system, consisting of devices exhibiting both electrical and thermal behavior. This paper describes several methods of modeling and simulating electrical and thermal effects in a single simulation. This is accomplished by implementing the electrical, thermal and electro-thermal models in an analog hardware description language (AHDL). The focus of this paper is on modeling the thermal and electro-thermal portions of the network including self-heating and heat transfer between junction and ambient. An automobile seat position controller circuit is described as an example of the application of these techniques.

4.1. INTRODUCTION

Although SPICE-based circuit simulators have traditionally simulated electronic circuits using detailed semiconductor device models, it is quite difficult to simulate the behavior of modern semiconductor devices that dissipate a considerable amount of heat. Besides the fact that the SPICE semiconductor models do not include some of the effects that

93

J.-M. Bergé et al. (eds.), Modeling in Analog Design, 93–120.

are important in power devices, such as ambipolar transport and high-level carrier injection, they also assume a constant operating temperature. In the typical approach, a user specifies the temperature for the device models prior to simulation, which remains constant during simulation. This temperature may be the same for all devices, or in some SPICE derivatives it can be different for each device [1]. Nevertheless, the temperature is held constant for each device. This treatment of thermal aspects is referred to as *static thermal* analysis. This approach has become increasingly inadequate for modern power electronic devices and other advanced semiconductors because of their considerable power dissipation. This power increases the internal device temperature (*self-heating*) as well as the temperature of other adjacent parts (*thermal coupling*).

This paper focuses on the modeling and simulation of thermal effects in electronic networks, including the interaction between the electrical and the thermal parts. This includes modeling of the thermal effects of chip, packages and heat sinks, the modeling of self-heating in device models, and the simulation aspects of electro-thermal networks. Although the consideration of thermal effects is usually only important in power applications, the paper will not address incorporating high-level carrier injection or nonquasi-static effects into semiconductor models to make them suitable for such applications.

Some results about electro-thermal modeling and simulation have previously been reported by Hefner and Blackburn [2,3]. This contribution builds on their work by taking a similar approach, but enhances it in several aspects, by presenting and comparing three methods of modeling the thermal aspects, and by presenting guidelines for the spacing of the grid used to discretize the heat diffusion equation. Further, the details of implementing self-heating effects in an existing model are described.

The paper first gives an introduction into electro-thermal simulation, by describing the basic mechanisms of heat transfer and of electro-thermal networks. Section 4.3 describes the implementation of self-heating in a MAST®[3] model, both conceptually and using a specific example. Section 4.4 presents and compares three methods to model the thermal aspects of chip, packages and heat sinks. Simulation considerations are covered in Section 4.5, and finally Section 4.6 presents an example where the thermal effects are essential for the operation of the circuit.

4.2. FUNDAMENTALS OF ELECTRO-THERMAL SIMULATION

The key to a successful electro-thermal simulation is the fact that the system to be simulated is a closed system, consisting of devices exhibiting both electrical and thermal behavior. Because it is a closed system, energy is conserved. In other words, the

---

[3] MAST® is a Registered Trademark of Analogy, Inc. SABER® is a Registered Trademark of American Airlines, Inc., licensed to Analogy, Inc.

electrical energy dissipated in the system is converted into thermal energy, or heat. Conversely, heat flow in the thermal portion of the system creates a temperature profile, which feeds back into the temperature dependent models used in the electrical portion. The purpose of electro-thermal simulation is then to find the equilibrium of the system, both steady state and dynamically.

There are three basic methods of heat transfer: conduction, convection and radiation. Conduction is the primary heat transfer mechanism in solid materials. It describes how thermal energy flows from the junction to the case and possibly to a heat sink. Convection-type heat transfer is based on movements in a medium, usually in liquids and gases. It describes, for instance, how the thermal energy is carried away from the heat sink. Radiation is the result of energy carried by electromagnetic waves. For the operating conditions of interest it is in the order of a few Watts per square meter and can be neglected.

4.2.1.   Heat Conduction

Heat conduction in a solid body is described mathematically by Fourier's law of heat conductivity

$$\vec{j}_Q = -k \cdot \nabla T \tag{1}$$

where $\vec{j}_Q$ is the heat flow density, k is the thermal conductivity, which depends on the material and possibly on temperature, and ∇T is the temperature gradient. Together with the continuity equation

$$\frac{\partial \rho_Q}{\partial t} + \nabla \vec{j}_Q = 0 \tag{2}$$

where ρ_Q is the heat density, we have, for a homogeneous and isotropic body,

$$\rho c \frac{\partial T}{\partial t} = \nabla \left(k(T) \nabla T \right) \tag{3}$$

where ρ is the material density, and c is its specific heat. Eq. (3) is called the *heat diffusion equation*; it is a partial differential equation. Section 4.4 will describe how this equation can be approximated under certain simplifying assumptions and implemented in a MAST model. Table 1 lists the material properties appearing in the diffusion equation for the relevant materials.

Material	Usage	ρ [g/cm^3]	c [J/gK]	k(T) [W/cmK]
Si	chip	2.328	0.7	$1.5486 \cdot (300/T)^{4/3}$
Cu	package	8.89	0.385	3.82
Al	heat sink	2.70	0.938	2.30
air	ambient	0.0129	1.006	0.00026—0.00035

Table 1:   Material Constants

4.2.2.   Heat Convection

Heat convection can be divided into natural convection and forced convection. If the movement of the medium (i.e., gas or liquid) is forced, we have forced convection. On the other hand, for natural convection the movement in the medium is caused by the density differences due to temperature differences. The mathematical description of convection is quite difficult, because it may involve a change from liquid to gaseous state of the medium, and laminar or turbulent flow. For our applications it is sufficient, though, to approximate heat convection by suitable thermal resistances. Between a heat sink and air the thermal resistance for natural convection is given by [6]

$$R_{nat} = \frac{1}{4.84 \cdot 10^{-4} A_{fin} \cdot \left(\dfrac{T_{fin} - T_{amb}}{h_{fin}} \right)^{0.35}} \tag{4}$$

where A_{fin} is the area of the heat sink fins, h_{fin} is their height, and T_{fin} and T_{amb} are fin temperature and ambient temperature, respectively. The thermal resistance for forced convection is

$$R_{for} = \frac{1}{\left(1.7 + f \cdot \ln \dfrac{v_{air}}{508} \right) \cdot 4.88 \cdot 10^{-4} A_{fin} \cdot \sqrt{\dfrac{v_{air}}{h_{fin}}}} \tag{5}$$

where v_{air} is the forced air velocity, $f = 0.148$ for $v_{air} \leq 508\,\text{cm/s}$ and $f = 0.433$ for $v_{air} > 508\,\text{cm/s}$.

4.2.3.   Electro-Thermal Networks

All materials exhibit some resistance to heat flow, and they also have a capacity to store heat, expressed by their specific heat. Therefore, thermal networks consist mainly of thermal resistances and thermal capacitances (representing chips, packages, heat sinks, and ambient), together with power and temperature sources. As described earlier, heat flow is equivalent to power, and the temperature profile is the result of heat flow in a thermal network. Energy conservation dictates that the sum of all power contributions at a thermal node equal zero, and that the temperature at all terminals connected to a thermal node beidentical. The notion of a conservative system as it relates to simulation can be generalized by referring to two entities: through variables and across variables. Through and across variables are important to understand because they indicate to the simulator the physical quantities to be conserved. An AHDL model associates a through and an across variable with each connection point (or node) in the model. In general, the through variables are summed to zero at a connection point as a means to insure a conservative system (a generalization of KCL). For electrical circuits current is the through variable and voltage is the across variable. In a thermal network power is the through variable and temperature is the across variable. The simulator, assuming a modified nodal formulation, imposes Kirchhoff's Current Law when solving a system of equations that represent an electrical circuit. This same condition is applied analogously to power for thermal systems. Thus, the simulator solves for the dynamic temperature distribution within the thermal network in the same way that node voltages are solved for within an electrical network. Due to this generalization of conservation, it is straightforward to implement both electrical and thermal systems simultaneously from a simulation standpoint.

In an electro-thermal network the power dissipated in a device is directly fed into the thermal network through a thermal pin. In other words, each dissipative device acts as a power source in the thermal network, with its positive pin connected to the thermal reference. In return, the temperature at the thermal pin of a device is the operating temperature of the device. This permits the device model to adjust its behavior as a function of its operating temperature (i.e., the model can support self-heating). The simulator now solves simultaneously for the dynamic temperature distribution within the thermal part of the system *and* the voltage/current distribution in its electrical part. This is known as *dynamic thermal* analysis.

Figure 1 shows how electro-thermal models for semiconductor devices couple electrical and thermal networks. The electrical terminals of these device models (e.g., IGBTs and power diodes) are connected to the electrical network and their thermal terminal is connected to the thermal network. The thermal network is represented as an interconnection of thermal components that allow a system designer to readily interchange different thermal components and examine different configurations of the thermal network. The thermal models for power modules and heat sinks can contain multiple terminals and account for the thermal coupling between the adjacent semiconductor devices.

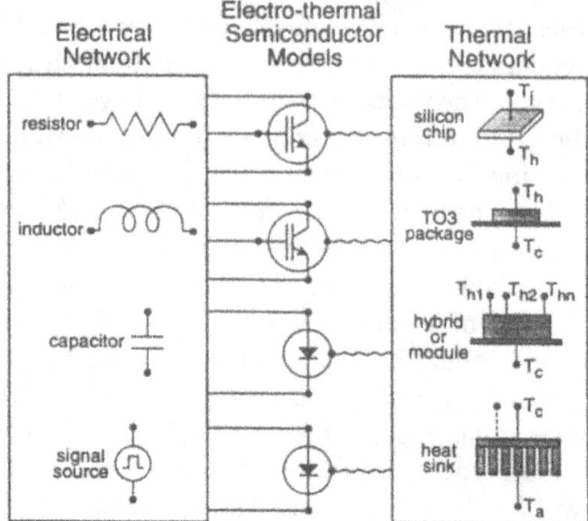

Figure 1:   Connection of electrical and thermal networks through electro-thermal models for semiconductor devices. (Reprinted with permission of National Institute of Standards and Technology)

One example of a simple electro-thermal network is shown in Figure 2, where an IGBT in a TO247 package has been mounted on a TTC1406 heat sink [2]. In addition to its electrical pins the IGBT has a thermal pin T_j representing its junction, which is connected to a thermal model of the chip. The other terminal of the chip model is connected to the TO247 package, which is itself connected to the heat sink. The heat sink finally is connected to a temperature source representing the ambient temperature.

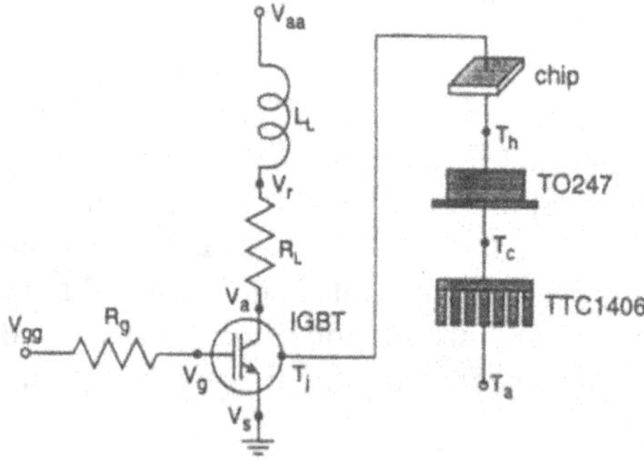

Figure 2:   Electro-thermal network for an IGBT with a TO247 package and TTC1406 heat sink. (Reprinted with permission of National Institute of Standards and Technology)

Figure 3 summarizes the concepts of electro-thermal modeling and simulation [3]. It shows the structure of the electro-thermal semiconductor models, indicating the interaction with the thermal and electrical networks through the respective electrical and thermal terminals. These models use the instantaneous device temperature (T_j at the silicon chip surface) to evaluate the temperature-dependent properties of silicon and the temperature-dependent model parameters. These values are then used by the physics-based semiconductor device model to describe the instantaneous electrical characteristics and instantaneous dissipated power. The dissipated power is calculated from the internal components of current, because a portion of the electrical power delivered to the device terminals is dissipative and the remainder charges the internal capacitances. The dissipated power calculated by the electrical model supplies heat to the surface of the silicon chip thermal model through the thermal terminal.

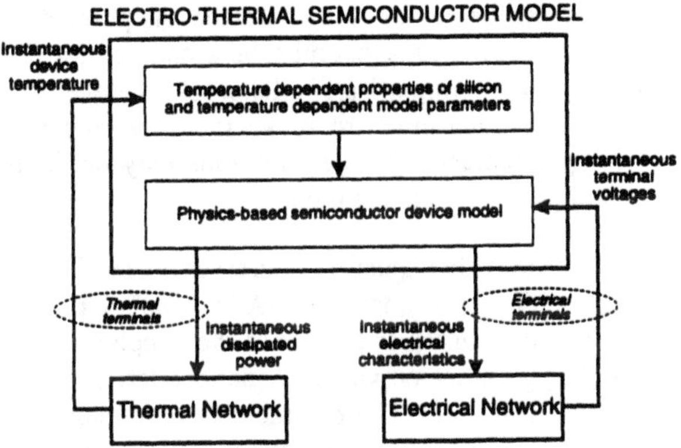

Figure 3:   Structure of electro-thermal semiconductor device models.(Reprinted with permission of National Institute of Standards and Technology)

4.3.   MODELING SELF-HEATING EFFECTS

As described in the previous section, in a network of thermal models temperature is a system variable that is solved for by the simulator. This is different from traditional circuit simulator semiconductor models in which temperature is a constant parameter specified by the user. Therefore, in order to implement dynamic electro-thermal effects in SPICE-based simulators, the first derivatives of the semiconductor model equations with respect to temperature would be required. The reason for this requirement is that SPICE needs continuous first derivatives for each model relationship with respect to all relevant system variables. Conversely, temperature effects are implemented into the Saber simulator by simply evaluating the temperature-dependent model equations in the MAST template.

The temperature dependent properties of silicon are well known; some are listed in Table 2. The temperature-dependent model parameters are obtained through experimentation by using the extracted values of the model parameters versus temperature. One advantage of using a physics-based model for the semiconductor devices is that the temperature-dependent properties of silicon can be used to describe the temperature dependence of the model, while only a few temperature-dependent model parameter expressions must be developed. Conversely, semiconductor models that rely heavily on empirical-based formulas require calibration of each of the empirical formulae versus temperature.

4.3.1. Implementing Self-Heating using the MAST Language

Implementing self-heating effects in a MAST-based model is straightforward. Since the Saber simulator does not require that first derivative information be specified in the model, the focus remains on utilizing or developing the proper temperature-dependent expressions that are required for the device being modeled. As pointed out above there are two fundamental types of relationships that must be accounted for. The first are the well-known temperature-dependent properties of silicon (or whatever material is being modeled). The second set of equations are those that describe the temperature dependence of the model parameters. These expressions may be difficult to derive depending on the nature and parameters of the model.

There are only a few basic items that must be addressed to implement self-heating effects in a semiconductor model. The simplified MAST template of Figure 4 outlines these items. They are described from top to bottom as the template is listed. However, the order of the sections (i.e., parameters, values, equations, etc.) is arbitrary. A brief description of these sections is in order. The parameters section contains all one-time calculations performed in a model. Common calculations that are performed here are the temperature relationships in a static thermal model. The governing equations of the model are implemented in the values section. This is where, for instance, the exponential current-voltage relation for a diode model is implemented. Lastly, the equations section describes the terminal characteristics of the model. The entities computed in the values section are combined to produce the total contribution to the terminals of the device being modeled.

Referring again to Figure 4, this *pseudo*-template outlines those additions/modifications that are necessary to convert a static thermal model to a dynamic (or self-heating) thermal model. First, a thermal pin is declared. This may be done with units of Kelvins, degrees Fahrenheit or degrees Celsius. While the units of degrees Celsius is the most common, typically the temperature must be converted to Kelvins internally for implementation purposes. However, the choice of units for the thermal pin is made based on input and output considerations.

```
template semi_therm a b c tj = argument1, argument2 # Template header
electrical a, b, c                            # Electrical nodes
thermal_c tj                                  # Thermal node
{                                             # Template body
# Local declarations
val tk tempj  # Junction temperature (Kelvins)
val p pwrd    # Total power dissipation
val ...       # Other local declarations as done without self-heating
parameters {  # Parameters section (one-time calcs.)
    #...Remove those items that are now functions of instantaneous
    #   temperature tempj and relocate them in the values section
    }
values {
    tempj = tc(tj) + 273.15          # Definition of junction
                                     # temperature in Kelvins
    #...Implement temperature-dependent properties of silicon
    #   (Table 2)
    #...Implement model parameter variation as a function of bias
    #   (as done in static thermal case)
    #...Implement model parameter variation as a function of tempj
    #...Implement governing equations
    #...Calculate power dissipation
    pwrd = i1**2 * r1 + i2**2 *r2 + . . .
    }
equations {
    #...Equations as implemented without self-heating...
    p(tj) -= pwrd            # Total power dissipation sourced
                             # out of thermal pin tj
    }
}
```

Figure 4: MAST outline form of a template with self-heating effects modeled.

The primary temperature that appears in all of the model relationships is the instantaneous junction temperature. This entity is declared in the local declarations section of the template (**tempj** in this case). Other local declarations have to do with the components of power dissipation and those entities that were formerly calculated in the parameters section (for the static temperature case) that are now functions of the instantaneous junction temperature **tempj**.

If converting a static thermal model to a dynamic thermal model, the parameters section is actually simplified since many one-time calculations now become iterative calculations that must appear in the values section.

In the values section **tempj** is first defined in Kelvins. The temperature-dependent properties of silicon such as those in Table 2 are then implemented. Next, the model parameters as a function of bias are calculated; then, these results are used to calculate the model parameters as a function of temperature. The governing equations of the model are implemented as before except that they now utilize the temperature-modified model parameters. Lastly, the dissipative portions of the governing equations are summed together to obtain a total power dissipation value. This value is subsequently "sourced" out of the thermal pin in the equations section, thus being the only modification to that section.

Mobility	$\mu(T_j) = \mu(T_0) \cdot \left(\dfrac{300}{T_j}\right)^{2.5}$
Diffusivity	$D(T_j) = \mu(T_j) \cdot \left(\dfrac{k \cdot T_j}{q}\right)$
Intrinsic Concentration	$n_i(T_j) = 3.88 \times 10^{16} \cdot \dfrac{(T_j)^{1.5}}{e^{\frac{7000}{T_j}}}$
Electron Saturation Velocity	$v_{nsat}(T_j) = 10^7 \cdot \left(\dfrac{300}{T_j}\right)^{0.87}$
Hole Saturation Velocity	$v_{psat}(T_j) = 8.37 \times 10^6 \cdot \left(\dfrac{300}{T_j}\right)^{0.52}$
Bandgap Voltage	$E_g(T_j) = 1.16 - \dfrac{\left(7.02 \times 10^{-4} \cdot T_j^2\right)}{T_j + 1108}$

Table 2:   Temperature-dependent properties of silicon

4.3.2.   Dynamic Thermal MOSFET

In this section an example of implementing self-heating effects is provided. The example is a simplified MOSFET model. This model is not intended to represent that of a state-of-the-art technology, but simply to convey the details necessary for one to follow to create a self-heating model using the MAST analog hardware description language (AHDL).

The MOSFET model implemented in this illustration is a level 1 model. It employs the Shichman-Hodges current equations and the charge formulation due to Sakallah, et. al. [4,5]. These equations are not repeated here or in the MAST listing since they are not the focus of this paper. In the implementation these relationships are accessed through MAST function calls to simplify the presentation of the template. The MAST listing of the template **mos** is given in Figure 5.

```
template mos d g s b tj = model, l, w
electrical d, g, s, b              # Electrical pin declarations
thermal_c tj                       # Thermal pin declaration
number    l = 1u,w = 1u            # Channel length, Channel width
mostype..model model=()            # Declaration of model by
                                   # reference to mostype
{
electrical dp, sp                  # Internal nodes
mostype..work work=()              # Declaration of work by
                                   # reference to mostype
val i      ids, ibd, ibs           # Branch currents
val q      qg, qb, qd, qs, qbsd, qbdd, qgso, qgdo, qgbo    # Charges
val tk     tempj                   # Junction temperature in Kelvins
val p      pwrd                    # Total power dissipation in Watts
parameters {
    work = mosparam(model,w,l)     # Geometry-dependent calculations
                                   # in MAST function

    }
values {
    # Definition of junction temperature in Kelvins
    tempj = tc(tj) + 273.15
    # Call mosval to get current and charges
    (ids, qg, qb, qd, qs) =
              mosval(work, model, v(g,sp), v(dp,sp), v(b,sp), tempj)
    # Call mosdiode to get currents and charges for bulk diodes
    (ibs, qbsd) = mosdiode(work, model, v(b,sp), tempj)
    (ibd, qbdd) = mosdiode(work, model, v(b,dp), tempj)
    qgbo = work->wcgbe*v(g,b)      # Calculate charges on overlap caps
    qgdo = work->wcgde*v(g,dp)
    qgso = work->wcgse*v(g,sp)
    # Calculate power dissipation
    pwrd = ids*v(dp,sp) + ibd*v(b,dp) + ibs*v(b,sp)
    if (model->rd ~= 0)      pwrd = pwrd + v(d,dp)**2/model->rd
    if (model->rs ~= 0)      pwrd = pwrd + v(s,sp)**2/model->rs
    }
control_section {
    if (model->rd == 0) .collapse(d, dp) # Collapse external and
                                   # internal drain when rd = 0
    if (model->rs == 0)  collapse(s, sp) # Collapse external and
                                   # internal source when rs = 0
    }
equations {
    i(dp->sp) += ids
    i(b->dp) += ibd + d_by_dt(qbdd)
    i(b->sp) += ibs + d_by_dt(qbsd)
    i(g->sp) += d_by_dt(qgso)
    i(g->b) += d_by_dt(qgbo)
    i(g->dp) += d_by_dt(qgdo)
    i(g) += d_by_dt(qg)
    i(b) += d_by_dt(qb)
    i(d) += d_by_dt(qd)
    i(s) += d_by_dt(qs)
    if(model->rd ~= 0)       i(d->dp) += v(d,dp)/model->rd
    if(model->rs ~= 0)       i(s->sp) += v(s,sp)/model->rs
    p(tj) -= pwrd
    }
}
```

Figure 5: MAST listing of simplified MOSFET model with self-heating (mos.sin).

This MOSFET model consists of five connection points. The first four are the normally thought of electrical connections (drain, gate source, and bulk, respectively). The next connection point is the thermal connection representing the junction temperature. It is declared as a thermal_c type connection point which indicates that it has the units of degrees C for temperature (the across variable) and Watts for power (the through variable). The other header declarations are those associated with the arguments of the template.

There are two internal electrical nodes that begin the local declarations section. All of the declarations contained before the parameters section are values that are needed in the values section of the template. Note that the junction temperature (in Kelvins) and the total power dissipation are additional values that are needed when adding self-heating effects to this model.

The parameters section is simply a call to the MAST function **mosparam** in which geometry-dependent variations of the model parameters are calculated. These values must not be temperature dependent. If they were they would have to be implemented in the values section in order for the results to be dependent upon the instantaneous junction temperature.

The values section first defines the junction temperature in terms of the connection point **tj**. The next several statements implement the governing equations of the MOSFET device. Note that **tempj** is passed to the MAST functions indicating that these relationships are a function of the instantaneous junction temperature. The temperature dependent properties of silicon and model parameter variation as a function of temperature are performed in these MAST functions. Finally, the total power dissipation is computed.

The equations section consists of statements indicating how the governing equations implemented in the values section are related to one another and to the connection points of the device. The electrical portion of these equations is the same as it would be without self-heating effects. The only change in this section is to source the dissipated power out of the thermal pin as the last statement indicates.

4.4.   THERMAL MODELS

In simulating an electro-thermal network, the thermal network is represented as an interconnection of thermal component models. As shown in Figure 2 each component represents an individual building block, for instance the chip, the package or the heat sink. The thermal models are typically lumped approximations of the heat flow through the chip, package, or heat sink material. The order of the approximation determines the accuracy achieved. In the following we will present three methods of creating such thermal models and compare them.

4.4.1. Thermal Models Based on Data Sheet Parameters

The data sheets for power devices usually include information about their thermal properties. However, this information is typically restricted to providing just the thermal resistances $R_{\Theta JA}$ between junction and ambient, and $R_{\Theta JC}$ between junction and case, respectively. Similarly, data sheets for heat sinks only specify, besides the dimensions of the sink, their thermal resistance $R_{\Theta CA}$ to air. This allows one to create a dc or static model for the thermal behavior, as shown in Figure 6.

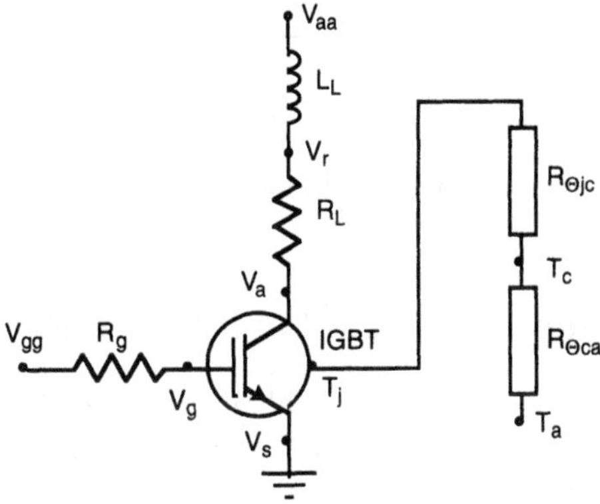

Figure 6: IGBT with static thermal model using data sheet parameters.

As with all static models, this simple model for the thermal behavior of the package and heat sink has its limitations. For instance, it can predict, given the average power dissipation of the device, what the average junction temperature will be. Alternatively, given the peak power, a pessimistic estimate of the peak junction temperature can be obtained. However, because no dynamic effects are included, the usability of this thermal model is limited to simple applications, where the power dissipation is more or less constant.

4.4.2. Lumped Models for Thermal Devices

A better thermal model can be created by using the package dimensions and material properties of the power device and the heat sink. The thermal capacity of each body is given by

$$C_{th} = volume \cdot \rho \cdot c \tag{6}$$

where ρ and c are as described in Section 4.2.1. Consequently, each of chip, package and heat sink are now represented by a thermal capacitance, with the thermal resistances in between. This arrangement is shown in Figure 7. Note that the thermal resistances have been split, and the thermal capacitances are connected to the thermal reference.

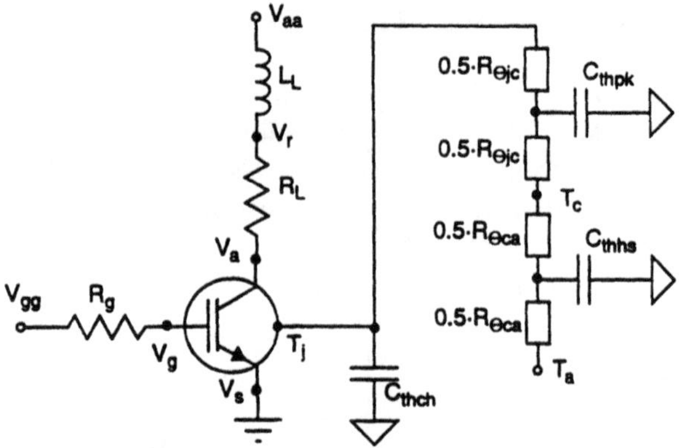

Figure 7: IGBT with lumped thermal models for chip, package and heat sink.

4.4.3.   Lumped Approximations for Diffusion Equation

The most accurate thermal model can be obtained by implementing the diffusion equation and convection. While implementing the convection equations (4) and (5) is straightforward, the diffusion equation is a partial differential equation, and PDEs are not supported in the MAST language. However, it is possible to use a lumped approximation to the diffusion equation, which only requires ordinary differential equations. We can even assume that the heat flow is one-dimensional, flowing through the chip and package perpendicular to the area, and radially away from the hot spot in the heat sink. This assumption simplifies the approximation and the model considerably. The following description is based on work published by Hefner and Blackburn [2].

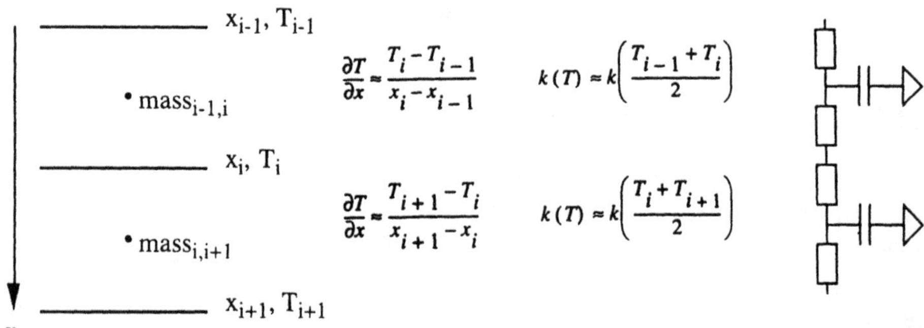

Figure 8: Discretization of diffusion equation, and thermal network equivalent.

For one-dimensional heat flow in a rectangular coordinate system the diffusion equation is given by

$$\rho c \frac{\partial T}{\partial t} = \frac{\partial}{\partial x}\left(k(T)\frac{\partial T}{\partial x}\right)$$

(7)

In order to discretize this equation, the volume carrying the heat flow is divided into n slices along the axis of heat flow, and each volume is represented by a single mass point. This process is schematically shown in Figure 8. In each of the regions the temperature gradient and the thermal conductivity are approximated by the expressions given in this figure. Inserting these expressions into Eq. (7) while discretizing again, we obtain

$$A \frac{x_{i+1} - x_{i-1}}{2} \rho c \frac{\partial T}{\partial t} = \frac{T_{i+1} - T_i}{R_{i,i+1}} - \frac{T_i - T_{i-1}}{R_{i-1,i}}$$

(8)

where A is the area of heat flow, and

$$R_{i,i+1} = \frac{x_{i+1} - x_i}{Ak\left(\dfrac{T_i + T_{i+1}}{2}\right)}$$

(9)

Eq. (8) says that the change of heat stored in a volume equals the difference of the heat flows on either side. This corresponds to the behavior of the thermal network shown on the right in Figure 8. Note that both thermal resistances and thermal capacitances depend on the material and the geometry of the bodies. Similar equations result for radial heat flow, using cylindrical coordinates.

$$\left[\left(\frac{r_{i+1} + r_i}{2}\right)^2 - \left(\frac{r_i + r_{i-1}}{2}\right)^2\right] \pi h \rho c \frac{\partial T}{\partial t} = \frac{T_{i+1} - T_i}{R_{i,i+1}} - \frac{T_i - T_{i-1}}{R_{i-1,i}}$$

(10)

$$R_{i,i+1} = \frac{\ln\left(r_{i+1} / r_i\right)}{2 \pi h k\left(\dfrac{T_i + T_{i+1}}{2}\right)}$$

(11)

where r is the radius and h is the height of the cylinder. The work in [2] also lists results for spherical coordinates.

The implementation of these equations in a MAST template is straightforward. Once the discretization grid has been selected the size of the slices and therefore the thermal capacitances are known and constant. The thermal resistances depend on temperature in the case of silicon, but are constant for copper and aluminum, the materials typically used for packages and heat sinks. As an example, Figure 9 shows the MAST implementation of the thermal model for a chip, using an equidistant grid with two internal nodes. The thermal capacities of the two slices are computed in the parameters section, and in the values section node temperatures in Kelvin and thermal resistances between nodes are determined. The equations section describes the heat flow between the nodes and the heat stored in the two slices.

```
template chip junct head = area, thick
thermal_c junct, head
number area=0.1            #...active area of chip (cm**2)
number thick=0.05          #...thickness of chip (cm)
{
#...Thermal constants for Silicon
number k0=1.5486           #...thermal conductivity at 300K (W/cm/K)
number kexp=4/3            #...k(T)=k0*(300/T)**kexp
number rho=2.328           #...density of silicon (g/cm**3)
number c=0.7               #...specific heat (J/g/K)
#...Discretization
number th1,th2             #...thickness of slices
number c1,c2               #...thermal capacity of slices
thermal_c tn1,tn2          #...internal thermal nodes (deg C)
val tk tj,t1,t2,th         #...temp. at junction, int.nodes and head (K)
val rth rj1,r12,r2h        #...thermal resistance between nodes
parameters {
     #...Compute thickness and thermal capacity of slices
     th1=thick/2 ; c1=th1*area*rho*c
     th2=thick/2 ; c2=th2*area*rho*c
     }
values {
     #...Compute temperatures in K
     tj=tc(junct)+273.15
     t1=tc(tn1) +273.15
     t2=tc(tn2) +273.15
     th=tc(head) +273.15
     #...Compute thermal resistances
     rj1=(((tj+t1)/(2*300))**kexp)/k0 * th1/(2*area)
     r12=(((t1+t2)/(2*300))**kexp)/k0 * (th1+th2)/(2*area)
     r2h=(((t2+th)/(2*300))**kexp)/k0 * th2/(2*area)
     }
equations {
     #...Heat flow between nodes
     p(junct->tn1) += (tj-t1)/rj1
     p(tn1->tn2) += (t1-t2)/r12
     p(tn2->head) += (t2-th)/r2h
     #...Heat stored in slices
     p(tn1) += d_by_dt(t1*c1)
     p(tn2) += d_by_dt(t2*c2)
     }
}
```

Figure 9: MAST implementation of thermal chip model.

The thermal models for package and heat sink are similar. The package model includes, as an added complexity, the two-dimensional lateral heat spreading, which has the effect that the effective heat flow area increases with depth into the package. The heat sink model describes the semi-radial heat flow from the package towards the heat sink fins, and natural and forced convection between the fins and the ambient [2].

One factor that greatly influences the accuracy of the approximation of the diffusion equation is the number and spacing of the discretization grid. While the example in Figure 9 uses a simple equidistant spacing, this arrangement does not give accurate results unless a large number of grid points is used. To understand why, let us review the dynamics in the thermal network. For high levels of power dissipation the heat is applied rapidly to the chip and diffuses only a few micrometers into the chip surface [2]. The chip-heating process is therefore non-quasi-static. The thermal network acts as a low-pass, filtering out the power peaks and responding with a smooth temperature profile. To reflect this properly in the thermal model, its input impedance should be as low as possible at high frequencies. For an RC-ladder low-pass filter this transforms into having the first resistance next to the source as small as possible, while the other resistances can be chosen larger. Consequently, the discretization should be selected such that the first grid point is quite close to the heat source, while the grid points farther away from the heat source can be spaced larger.

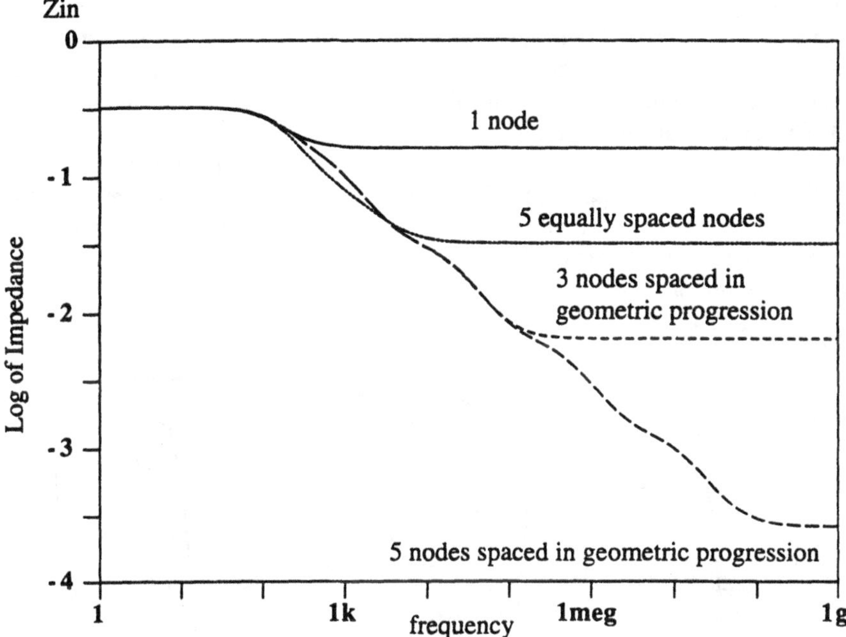

Figure 10: Input impedance of thermal model for chip for different spacing and number of nodes (grid points).

Figure 10 shows the thermal input impedance of the chip model for 1 grid point, 3 grid points with geometrically progressing spacing, and 5 grid points with both equal and geometrically progressing spacing. The figure shows that at high frequencies the input impedance equals the first resistance, and that the time constant of the first RC section corresponds to the 3 dB frequency of the impedance with respect to its high frequency constant value. It is clear from this graph that equal spacing of the grid points is inferior at high frequencies to spacing with a geometrical progression, and that for lower frequencies both the spacing and the number of grid points are almost irrelevant. The conclusion is that the number of grid points required to get an accurate thermal model depends on the spectrum of the power dissipation, and that for a given number of grid points a geometrical progression should be used for the spacing.

This conclusion has been confirmed by experiments with an IGBT circuit. The results obtained with a chip model consisting of five grid points with geometrically progressing spacing are at least as accurate as the ones reported in [2], where three logarithmically spaced banks of five linearly spaced grid points were used. A grid of three nodes still gives reasonably accurate results, except at very fast transients of the power dissipation.

4.4.4.   Comparison of the Three Methods

In order to compare the three methods we used a MOSFET in a source follower configuration. Using data sheet parameters for the TO220 package and for a commercial heat sink we developed models for the thermal properties of chip, package and heat sink. The static model, using method 4.4.1, consisted of the two thermal resistances from junction to case (including the chip and the package) and from case to ambient (the heat sink). The lumped model according to method 4.4.2 included two resistances and a capacitance for each the chip, the package and the heat sink. The method 4.4.3 model approximating the diffusion equation used 3 nodes spaced with geometric progression for both the chip and the package, and 10 nodes for the heat sink. The heat sink model required some minor adjustments, due to boundary effects, to exhibit the same dc behavior as the simpler models.

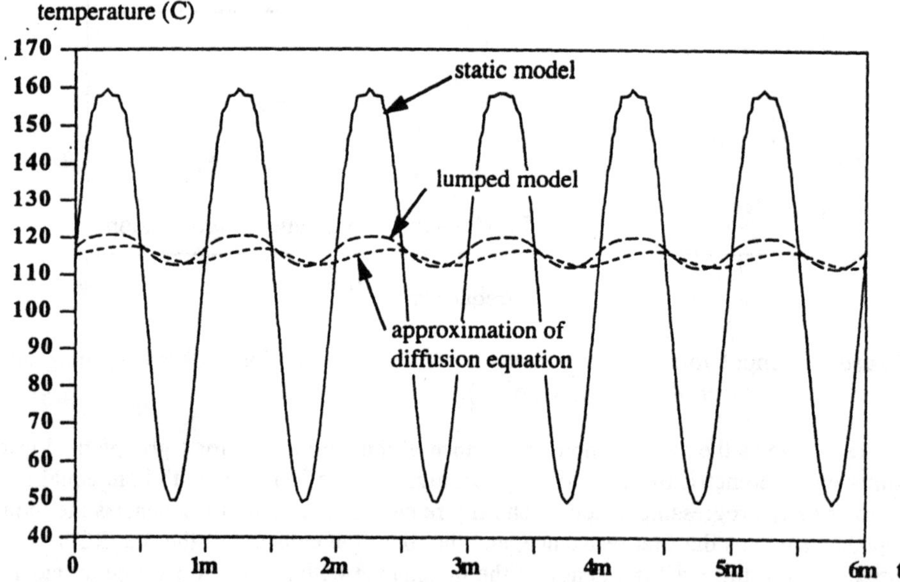

Figure 11:   Junction temperature of MOSFET using 3 different thermal models.

The source follower was then excited with a sinusoid of 1 kHz, applied in its active region. Figure 11 shows the simulation results. As expected, the junction temperature follows exactly the power dissipation when the static model for the thermal parts is used. Both the lumped model and the approximation of the diffusion equation provide much less pessimistic results, damping the junction temperature variations considerably. Because of its more accurate approximation of the diffusion effects, the model according to method 4.4.3 predicts the smallest variations for the junction temperature.

This example shows that for an accurate prediction of the behavior of a power device the thermal network must include models for the thermal storage effects. As described in the previous section, the number of grid points (i.e. whether to use methods 4.4.2 or 4.4.3) depends on the application, with faster power variations requiring more grid points.

4.4.5. Limitations of the Three Methods

As may have become apparent from the description in the previous sections, the three methods presented here only address one-dimensional heat flow. As such, the methods are well suited for power applications, where the thermal coupling between adjacent devices is relatively small because the devices are on different chips and even in different packages. Specifically, the methods address the case of a single heat source on a chip, where the lateral heat flow in the chip can be neglected. Consequently, the whole chip is considered to be at the same temperature. Further, the thermal models of the package and the heat sink used in the examples assume only a single heat source (more exactly, one heat producing area) per package and per heat sink, although it is possible to combine several adjacent heat sources into one if they are not too far apart, by adjusting the heat producing area appropriately.

In order to address chips with many heat sources, it is necessary to model the lateral heat flow in the chip, by creating a two-dimensional model of the thermal properties of the chip. Additionally, if the thermal properties of the package are important, the thermal model of the chip becomes three-dimensional. The thermal model of the package becomes more complex, too, if the value of its thermal capacity is comparable to the thermal capacity of the chip, or if there are multiple chips in the package. Similarly, the complexity of the heat sink model increases if there are multiple packages per heat sink. Addressing these issues is beyond the scope of this contribution.

4.5.   SIMULATION CONSIDERATIONS

Although the equations describing the thermal characteristics of devices such as chips, packages and heat sinks use different physical units than the equations describing electrical characteristics, they are analogous. For instance, in the thermal network the sum of heat flows entering a thermal node equals zero. Hence, any simulator supporting the description and simulation of electrical behavior can be used, in theory, to simulate thermal behavior and electro-thermal interaction. In other words, the issue is not the simulator (with some exceptions described below), it is the models. However, using an

AHDL to describe both the electrical and thermal properties of parts makes it particularly simple to first describe and then simulate electro-thermal interaction.

The simulation of electro-thermal networks poses some challenges to the analog simulation engine. Usually, the time constants in the thermal network (i.e. for heat flow within the silicon chip, package, and heat sink) are many orders of magnitude longer than the time constants in the electrical network. This means that the system of simultaneous equations to be solved by the simulator is usually stiff. As a consequence, an integration method suited for stiff systems must be used for the simulation of electro-thermal networks, for instance second order Gear.

Because of the wide spread of time constants self-heating effects behave dynamically. This is true even for circuit conditions that are considered to be steady-state for the electronic devices. As a consequence, electro-thermal simulations must usually be run over a longer time interval than traditional electrical simulations, and the time interval tends to increase for more accurate thermal models of chip, package and heat sink. Still, the simulation end time is mostly dependent on the operation of the circuit, and the impact of choosing better models for the thermal devices is relatively small. A method for reducing the time required to simulate electro-thermal systems at steady-state is described in [7].

One problem that may occasionally occur is that the simulator may find negative values for the absolute temperature. Typically this is caused by errors somewhere in a model, coupled with the fact that the simulator has no knowledge about the special nature of the absolute temperature. Although it is possible to add conditions to thermal models that force the temperature to stay positive, it is usually not worth the effort and has therefore been left out in the examples.

4.6.   APPLICATION EXAMPLE

In the past, electrical engineers would design and validate their circuits using a general electrical simulation tool, or some other software tool that was used only for a specific part of the design. Mechanical, thermal, and digital engineers would also use their specific software tools for validation of their designs. It was not easy to link these designs into one simulation and validate compatibility of the entire system. This application illustrates a practical engineering design problem incorporating thermal protection in an electro-mechanical motor drive system.

The seat position motor drive circuit is shown in Figure 12. This circuit contains six models which provide an integrated thermal, electrical, and mechanical simulation validation of the total design. The automobile battery powers the system when the switch is closed. The switch represents the button on the side of the car seat that activates the seat position motor. Once activated the motor drives the gears, which in turn drive a rack and pinion system within mechanical limits. The temperature-limited MOSFET is in series with the armature winding of the motor to provide thermal protection. This thermal protection circuit uses an electro-thermal model of a power MOSFET with temperature sense and shutdown circuitry.

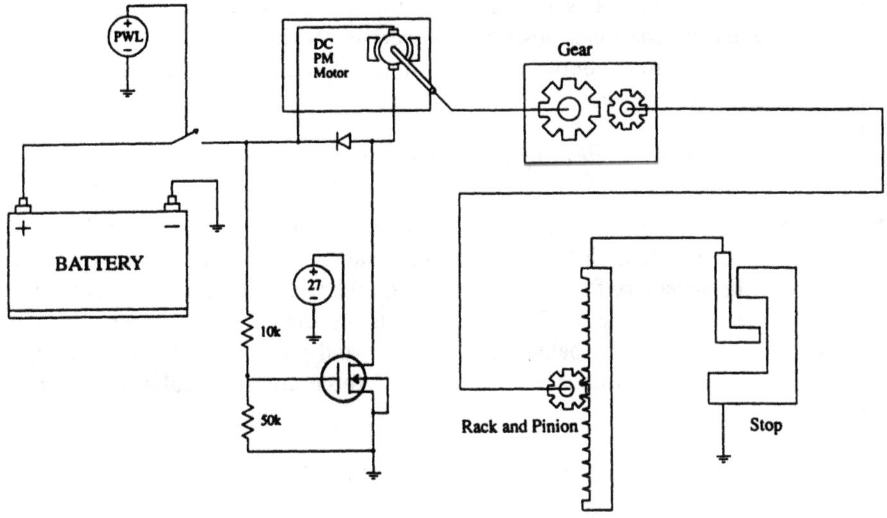

Figure 12:   Seat position motor drive with active thermal cut-out.

4.6.1.   **Mechanical Models**

The motor is modeled as a permanent magnet DC machine that may act as either a motor or generator depending on how it is connected. Machine parameters include the resistance and inductance of the armature, interpole and compensating windings. The torque and back emf constant are numerically equal to each other and are modeled with a single parameter. Internal damping and inertia of the unloaded machine as well as the number of machine poles are additional parameters.

The gear model represents a simple mechanical gear. Its behavior is such that a fixed relationship is maintained between the shaft 1 and shaft 2 angular velocities. That is, the velocity of shaft 2 is equal to the velocity of shaft 1, scaled by the gear ratio **gr** as in Eq. (12):

$$velocity\left(shft2 \right) = velocity\left(shft1 \right) \cdot gr \tag{12}$$

This velocity relationship is maintained by the model by generating the necessary torques on the two shafts. Power is conserved by the gear model, as the shaft torques are also scaled by the gear ratio

$$torque\left(shft1 \right) = -\left(torque\left(shft2 \right) \cdot gr \right) \tag{13}$$

Note the minus sign in the torque equation. This is because the reaction torque applied back on the *drive* shaft is opposite to the torque applied by the *driven* shaft.

The rack-and-pinion model acts as a rotational/translational converter, with mechanical rotation angle and translational position connections. Its behavior is such that a fixed relationship is maintained between the pinion rotation angle (*ang*) and the rack translational position (*pos*) as

$$radius \cdot angle\left(ang\right) = position\left(pos\right) - pos0 \qquad (14)$$

The scale factor *radius* which represents the radius of the pinion, and *pos0* is the zero angle position offset. The defined position/angle relationship is maintained by the model by generating the necessary forces and torques (on the rack-and-pinion, respectively) to enforce the relationship. The products of torque times angular displacement, and force times linear displacement, are balanced by the rack-and-pinion model (so that energy is conserved). This occurs because the torque and force are also scaled by the factor *radius*.

$$torque\left(ang\right) = -\left(radius \cdot force\left(pos\right)\right) \qquad (15)$$

Again, note the minus sign in this equation. This is because the reaction force (torque) that is applied back on the *driving* member is opposite to the torque (force) applied by the *driven* member.

The mechanical stop model represents a two position hardstop (mechanical limit). Its behavior is such that when the relative position (pos1 - pos2) is between the limits *ptop* and *pbot*, the model exerts no force. However, when the position exceeds either limit (that is, if the relative position increases above *ptop* or drops below *pbot*), then a force proportional to this excess is generated. The force polarity is such that the relative position is driven back into the non-limiting region. The proportionality constants (i.e., the mechanical "stiffness") for the top and bottom hardstop limits are given by the parameters *ktop* and *kbot*, respectively. In addition, an energy dissipation factor is available, to model the inelastic effects of collisions with the hardstop. The damping factors, *dtop* and *dbot*, provide this energy loss effect.

4.6.2.   Modeling the Temperature Limited MOSFET

The MOSFET in Figure 12 actually represents a MOS device along with temperature sense and shutdown circuitry. A block diagram of the temperature limited MOSFET is given in Figure 13. The circuitry utilizes the temperature effects on internal threshold voltage (V_T) and forward diode voltage drop (V_d) to provide the triggering mechanism versus temperature. Positive feedback as well as a hysteresis loop is also included in the circuitry. Figure 14 shows these characteristics of the MOSFET model. This figure shows the junction temperature of the device being forced to increase from 25°C to 200°C. The results show that the device quickly shuts off at approximately 175°C and then turns back on at approximately 135°C. This validates the hysteresis loop, the positive feedback, and the temperature triggering circuit of the MOSFET model.

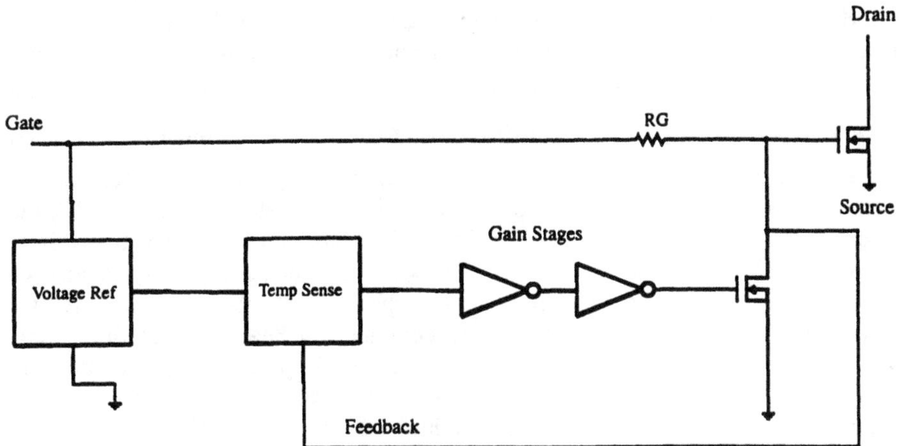

Figure 13:   Block diagram of temperature limited MOSFET device.

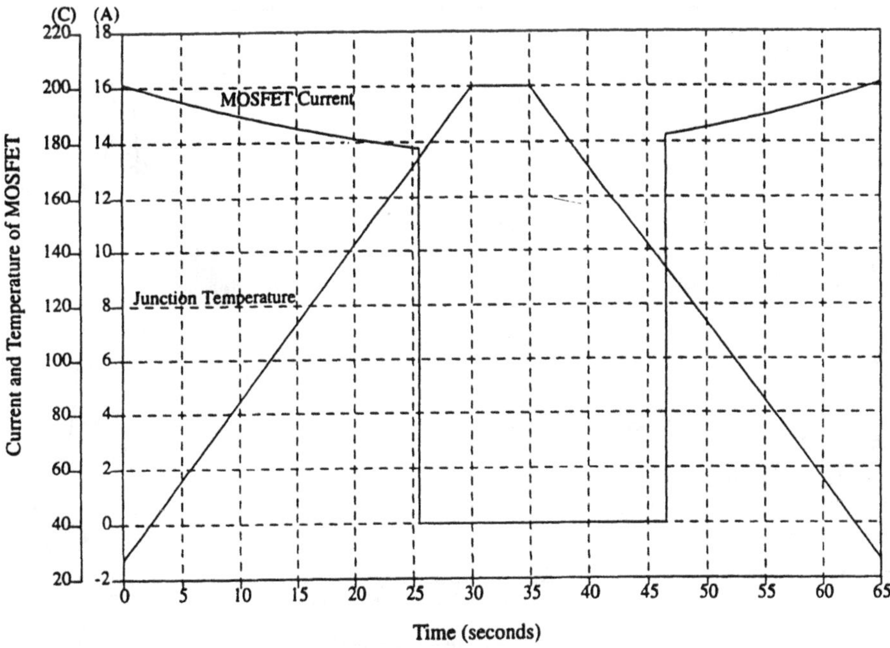

Figure 14: Junction temperature and current of temperature-limited MOSFET illustrating thermal hysteresis.

Figure 15 shows the junction temperature of the MOSFET model as the gate to source voltage is ramped from 20 volts to 0 volts. This simulation was performed without thermal models for the chip, package and heat sink attached. A first-order thermal network, a T-configuration similar to Figure 7, from junction to ambient was employed. The temperature rises dramatically as the device model traverses from it's linear region $\left(V_{gs} - V_T > V_{ds} \right)$, through the saturated $\left(V_{gs} - V_T < V_{ds} \right)$ and sub-threshold region $\left(V_{gs} < V_T \right)$, and finally into the cutoff region. As the device begins to shut off, the voltage across the device begins to increase and the power dissipation increases. This causes the device temperature to increase. Once the temperature reaches 175°C, the shutdown circuitry shuts the device off. However, the time constant associated with the thermal network is much faster than the rate at which the device is being controlled via the gate voltage. Thus, when the temperature drops to 135°C, the shutdown circuitry reactivates the MOSFET and it begins to conduct and heat up once more. This oscillation behavior continues until the gate voltage drops too low to sustain any appreciable current in the device. The envelope of the thermal oscillations does not remain fixed about 135°C and 175°C, but also increases (see Figure 15) since the model of the triggering mechanism in the shutdown circuitry of Figure 13 also possesses dynamic thermal effects.

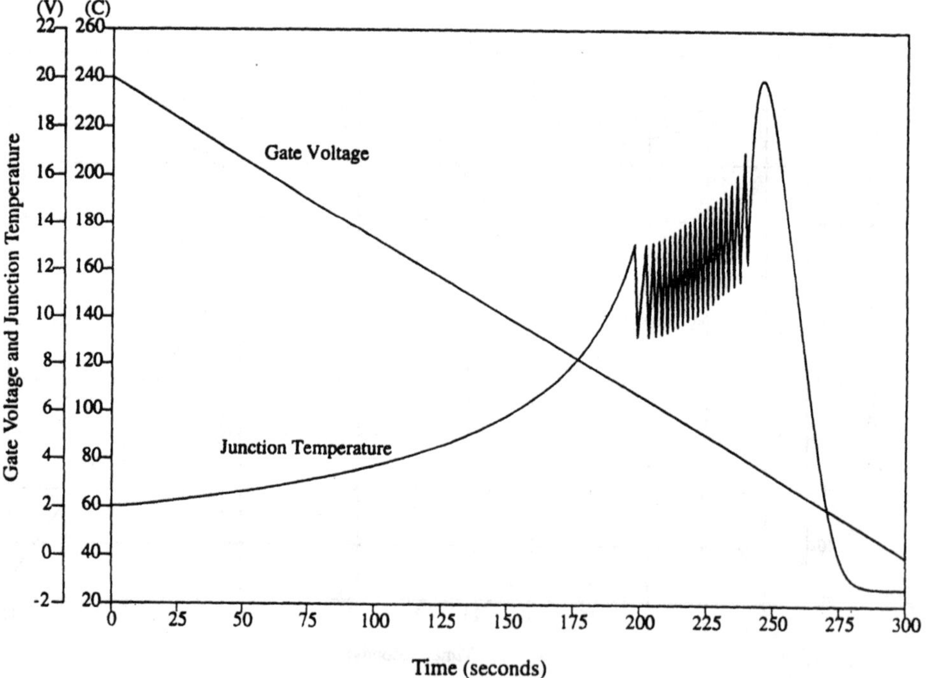

Figure 15: Junction temperature of MOSFET as gate voltage is slowly ramped downward illustrating thermal oscillations.

Based on the junction temperature profile observed for a particular load scenario, the designer may decide on using a certain package and/or heat sink for the MOSFET. Figure 16 shows the thermal gradient through the package and out to the heat sink as the gate to source voltage is ramped from 20 volts to 0 volts as before. The thermal waveforms include the junction temperature of the MOSFET device model, the temperature at the chip level, the package level, and finally at the heat sink. With this choice of heat sink the thermal oscillations are no longer observed. The thermal model of the package and heat sink can be altered to provide different thermal time constants representative of different package and heat sink configurations and materials.

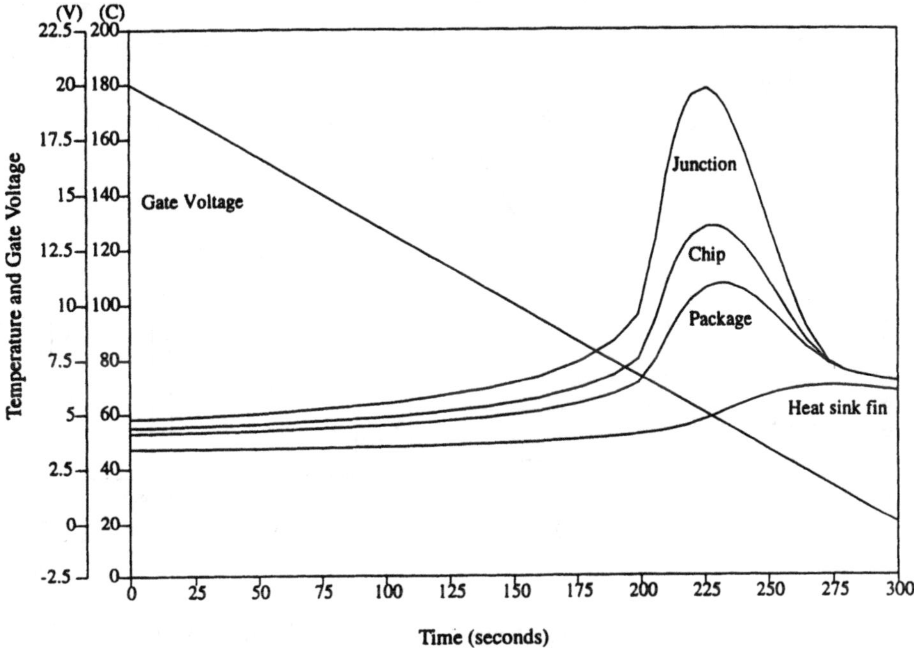

Figure 16: Thermal waveforms for the temperature limited MOSFET with heat sink as gate voltage is slowly ramped. Note the elimination of thermal oscillations.

4.6.3.   Integrated Seat Position System Simulation

Figure 17 shows the system level results of all the thermal, electrical, and mechanical models working together. The simulation objective is to activate the seat position motor and move the position of the seat until the seat hits the physical stops.

Normally the person holding down the button for the seat to move would release the button at this time, and the motor would shut off. However, the simulation assumes that a fault occurs, or perhaps a child is in the car playing with the seat position and does not release the button when the seat hits the physical stops. With the seat up against the stops and the motor button still engaged, the motor is forced into a stall condition. At this point the motor's armature current increases. The temperature limited MOSFET device model is in series with the armature winding, and therefore carries this increased current. This causes the MOSFET to heat up until the junction temperature rises to 175°C. At this temperature the MOSFET circuit shuts the device down and thus the motor shuts off. With no current flow through the MOSFET, it begins to cool down. If the seat position button is still activated when the MOSFET device model cools down to 135°C, the MOSFET will turn back on and the cycle will repeat.

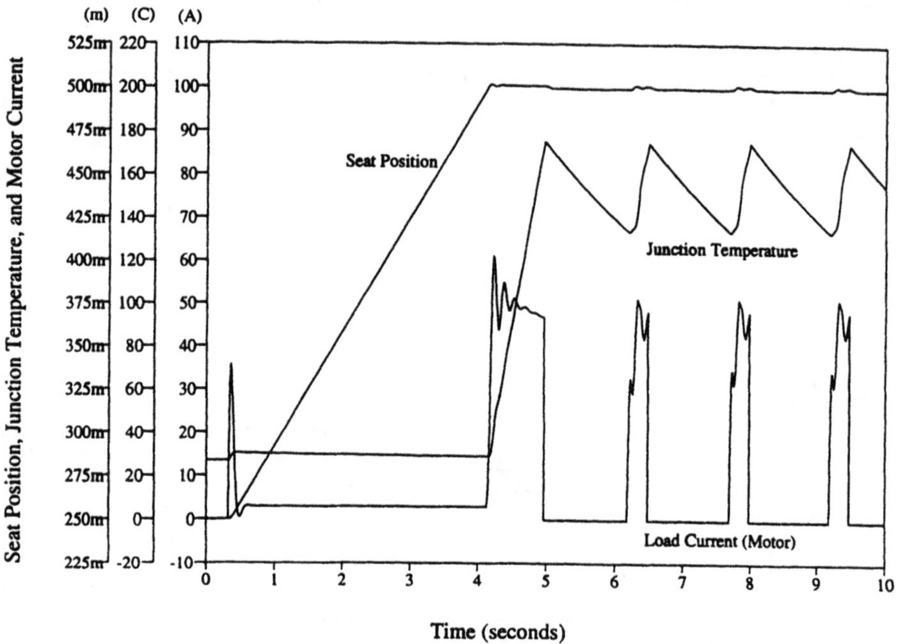

Figure 17: Seat position, junction temperature and motor current of seat position system.

4.7. CONCLUSIONS

In this paper we have shown that an analog hardware description language makes it easy to extend the simulation model of the system to include non-electrical behavior, using as an example the modeling of thermal aspects. This investigation was initiated by the increased interest in power applications, where the thermal aspects of the devices and their environment are essential for an accurate simulation of their operation.

We have summarized the fundamentals of heat transfer, including diffusion, convection and radiation, and have shown how these concepts can be represented in an electro-thermal network. We described how the partial differential equation describing the diffusion effects can be approximated using ordinary differential equations. Based on experiments, we demonstrated that the discretization grid should be selected such that the slice next to the heat source is small compared to the other slices, in order to get good high frequency performance. A grid spacing with geometric progression appears to be well suited. Using a discrete approximation of the diffusion equation and two simpler methods of creating thermal models we demonstrated that for accurate predictions of the thermal behavior it is essential to include the thermal storage effects.

We have explained how a standard device model can be extended to include self-heating effects, in order to be able to study thermal phenomena such as thermal oscillations. This process is relatively simple if physics based models are used for the devices, because the temperature dependence of the material can be easily incorporated in the device model. It is also much simpler to develop a model with self-heating in an AHDL than with a SPICE like simulator, because the models do not have to include derivatives of the behavior equations with respect to temperature. Finally, we have shown, using a mixed-technology example, that the modeling and simulation of electrical and thermal interaction is essential for the operation of some systems.

There are two places where further work is required to address the needs of power applications. Firstly, the device models must be enhanced to include high-level carrier injection and other effects that are important in power devices. Secondly, generic models for chip, packages and heat sinks should be developed, based on the approach presented in this paper, that are easily parameterizable such that they can be characterized for different commercial packages and heat sinks.

REFERENCES

[1]   L. W. Nagel, "SPICE2: A computer program to simulate semiconductor circuits," ERL Memo No. ERL-M520, University of California, Berkeley, 1975.

[2]   A. R. Hefner and D. L. Blackburn, "Thermal Component Models for Electro-Thermal Network Simulation," Proc. IEEE Semiconductor Thermal Measurement and Management SEMI-THERM Symposium, p. 88, 1993.

[3]   A. R. Hefner and D. L. Blackburn, "Simulating the Dynamic Electro-Thermal Behavior of Power Electronic Circuits and Systems," Conf. Rec. IEEE Workshop on Computers in Power Electronics, p. 143, 1992.

[4]   H. Shichman and D. A. Hodges, "Modeling and simulation of insulated-gate field-effect transistor switching circuits," IEEE J. Solid-State Circuits, SC-3, 1968.

[5]   K. A. Sakallah, Y. T. Yen, S. S. Greenberg, "A first-order charge conserving MOS capacitance model," IEEE Trans. on Computer-Aided Design, vol. 9, pp. 99-108, Jan. 1990.

[6]   G. N. Ellison, Thermal Computations for Electronic Equipment, Van Nostrand Reinhold, 1984.

[7]   H. A. Mantooth and A. R. Hefner, "Electro-thermal simulation of an IGBT PWM inverter," IEEE PESC Conf. Rec., pp. 75-84, 1993.

5

MODELING OF POWER MOSFET AND BIPOLAR TRANSISTORS TAKING INTO ACCOUNT THE THERMOELECTRICAL INTERACTIONS

C. Lallement, R. Bouchakour, T. Maurel

Département d' Electronique, ura CNRS 820, Telecom-Paris, 46 rue Barrault, 75634 Paris Cédex 13, FRANCE

ABSTRACT

We have developed analytical one-dimensional thermoelectrical models for the power mosfet and bipolar transistors, implemented in the Saber circuit simulator, in which the device's temperature becomes an interactive variable during the simulation. The accuracy of the models is appreciated with electrical and thermal characterizations, and validation circuits.

J.-M. Bergé et al. (eds.), Modeling in Analog Design, 121–143.

5.1.   INTRODUCTION

Power electronics devices dissipate a considerable amount of heat and are often operated in high-temperature environments; this directly influences both their static and dynamic operational characteristics, and their associated failure rates. The thermal aspects of circuit design can therefore be considered as having equal importance as the strictly electrical aspects. The device standard models in the currently analog simulators generally include temperature dependence, but it must be chosen by the user before the simulation and hence must remain constant at the predetermined value during the simulation, so the scope for the C.A.D. can be severely limited. Sarting a few years ago, we have seen device models [1 - 4] in which the chip temperature became a dynamic variable, and in which the electrothermal modeling philosophy differed only by the method used to calculate the new value of the chip temperature at each simulation step. In this context, this article tries to propose solutions to take the thermoelectrical interactions in the power mosfet and bipolar transistors into account. This solution associates the electrical model of the device to a thermal network, which models the different material layers crossed by the heat flow from the silicon chip to the ambient air and thus allows a determination of the chip temperature during the simulation. The accuracy of each model is appreciated with an electrical and thermal characterization, and a validation circuit.

5.2.   COMPLETE MODEL

The models' equations and topologies have been implemented into the Saber circuit simulator using the MAST description language. They are based on the interaction between an electrical network and a thermal network [5] coupled by the temperature(T) and the instantaneous dissipated power (P). The nature of the thermal network used determines the size of the T and P variables, responsible for the electrothermal interactions:

For a one-dimensional thermal solution (Figure 1-a), a uniform chip temperature is assumed, so T and P are scalar. T is introduced in the model as a chip temperature variation ΔTchip.

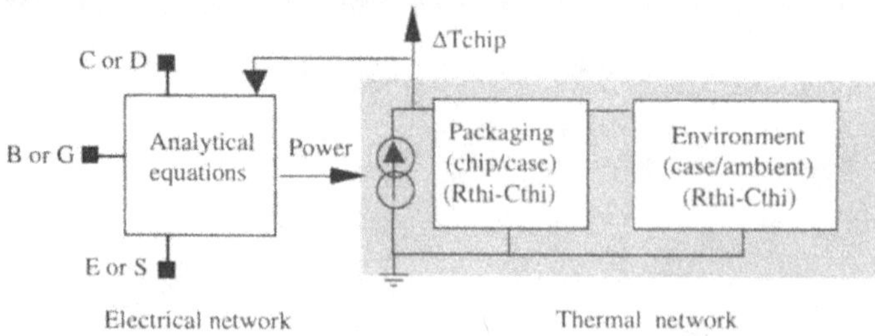

Figure 1-a: Complete one-dimensional model of the power transistors

For a more precise evaluation of the temperature gradient in the bipolar structure, a two-dimensional thermal model is used for the chip (Figure 1-b). It provides a set of local temperatures (so T is a vector) and needs P to be an instantaneous dissipated power profile.

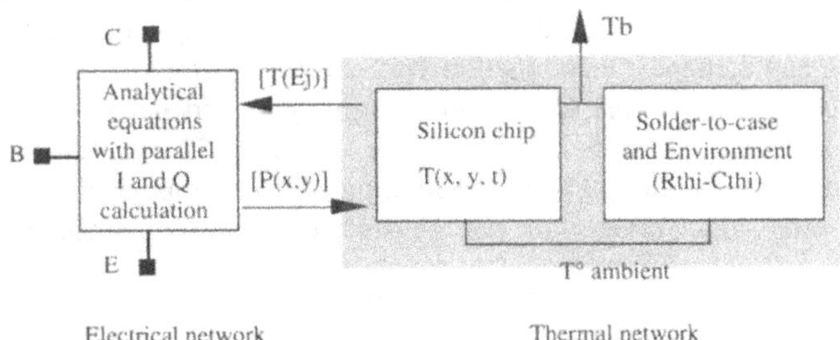

Figure1-b: Complete model of the power bipolar transistor with the 2D thermal network

5.3.   THE ELECTRICAL NETWORK

5.3.1.   The Power Mosfet Transistor (VDMOS Transistor)

The VDMOS transistor can be modeled [5-7] as an intrinsic MOS transistor (Spice level 3) representing the behavior of the channel; to this transistor are connected the CGD capacitor [8] and the access resistor to the channel Rd1 (which take the effects of the vertical structure into account), the drain resistor Rd2, the gate resistor Rg, the "freewheel" diode Dbody, and the source resistor Rs and inductance Ls. The electrical network (Figure 2) allows an analytical modelling of all intrinsic elements, and the insertion of a parameters calculation dependant on temperature at each simulation step.

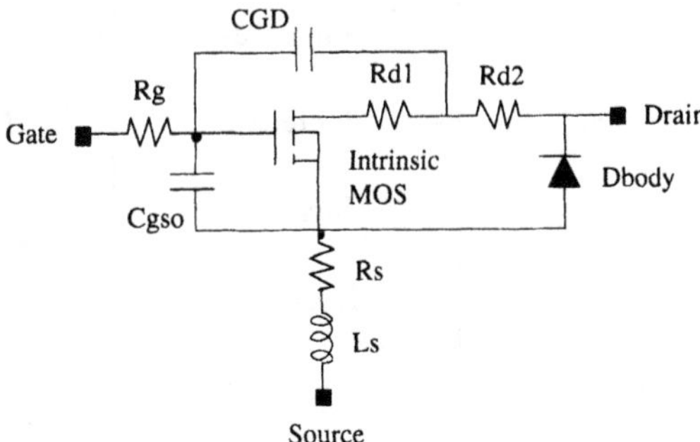

Figure 2:   Electrical model of the power mosfet transistor, all elements are sensitive to temperature except the Cgso capacitance and the Ls inductance

5.3.2.   The Power Bipolar Transistor

The electrical model (Figure 3) [9] contains an intrinsic bipolar transistor (standard Gummel-Poon model) [10, 11] to which are added elements that take the effect of the epitaxial layer into account. The semiconductor one-dimensional ambipolar equations allow an evaluation of the minority carrier profile in the high-injected lightly doped collector layer, and then the recombination base-collector current Irec, the voltage drop Vepi along the epitaxial layer, the stored charge Qepi and its derivative that both dominate the static and dynamic behavior of the transistor. The non-quasi-static effect is taken into account by a step-by-step evaluation of the ambipolar diffusion length λ with its static value, Qepi, $\partial(Qepi)/\partial t$, and the base-collector lifetime τbc [9].

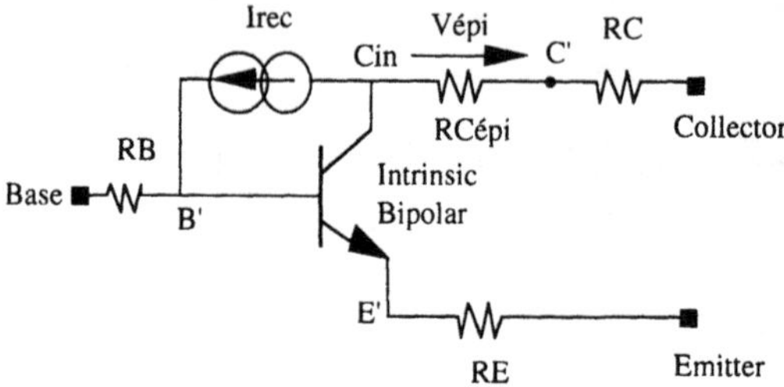

Figure 3:   Electrical model of the power bipolar transistor, all elements are sensitive to of temperature

Characteristics	Bipolar	Mosfet
Forward voltage drop	Reduces	Increases
Rise time	Increases	No change
Storage time, tur-noff delay time	Increases	No change
Fall time	Increases	No change
Vbe, threshold voltage (VT)	Decreases	Decreases
Gain, transconductance	Increases	Decreases

Table 1:   Effect produced by a temperature increase

5.4. THERMAL INFLUENCE ON THE ELECTRICAL BEHAVIOR OF THE TRANSISTORS

A comparison of the temperature dependence on some important mosfet and bipolar parameters is shown in Table 1.

5.4.1. The Power Mosfet Transistor

Several main effects are responsible for the variation of the thermal coefficient, as the $\Delta IDS/\Delta Tchip$ demonstrates by the static output characteristic (Figure 4).

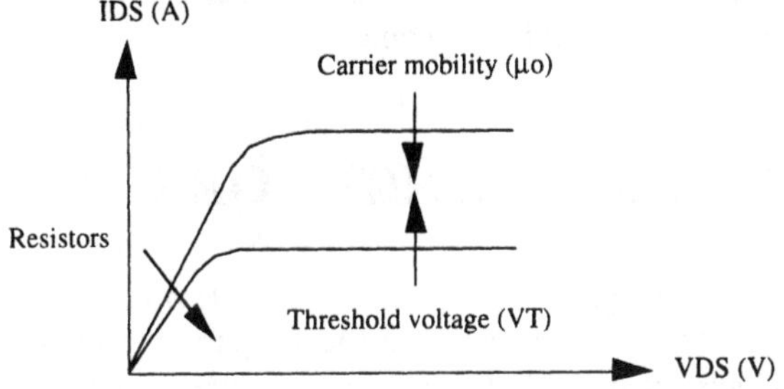

Figure 4: Temperature influence on the static characteristic : IDS = f (VDS)

The Temperature influences [11-13] the threshold voltage, the carrier's mobility in the channel, the mobility reduction factor due to the longitudinal electrical field, the intrinsic resistors, the intrinsic diode current and capacitance. An uniform chip temperature is assumed for the power MOSFET transistor.

5.4.2. The Power Bipolar Transistor

In the bipolar transistors, the thermal runaway and current-focusing are responsible for second breakdown. The IC current variations are dominated by several simultaneous phenomena (Figure 5).

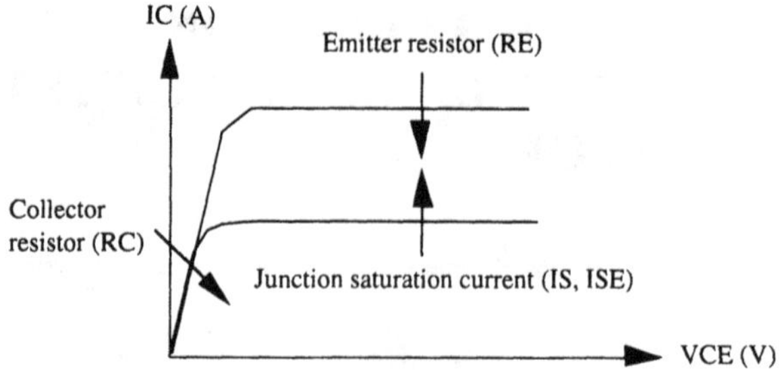

Figure 5: Temperature influence on the static characteristic : IC = f (VCE)

The main electrical parameters sensitive to temperature are the intrinsic resistors, the junction saturation current, the junction capacitances and the carrier's mobility. In the first thermal study, a uniform chip temperature is assumed, then a solution is used to take the chip temperature gradient into account.

5.5.   THE THERMAL NETWORK

To describe the heat propagation between the chip (primary heat source) and the environment, we use the analogy between the propagation of a heat flow and the propagation of an electrical signal [14]. In both cases, we can talk about resistances and capacitances (Table 2). Each material layer crossed by the heat flow created by the electrical power dissipated is equivalent to a {Rth(i) - Cth(i)} cell of the network.

Quantity	Electrical	Thermal
Temperature variation	V(V)	$\Delta T(K)$
Dissipation	I (A)	P(W)
Thermal resistance	$R(\Omega)$	Rth(K/W)
Thermal capacitance	C(F)	Cth(J/K)
Thermal conductivity		Kth(W/cm.K)
Material density		$r(g/cm^3)$
Specific heat		Cp(J/g.K)

Table 2:   Electrical-thermal analogy

5.5.1.   The Transistor Package (Chip-to-Case)

The number of cells used depends on the transistor package (Figure 6). The thermal network allows a calculation of chip temperature (T°chip) given by:

$$T°chip = T°ambient + Zth * Power = T°ambient + \Delta Tchip \qquad (1)$$

where Zth is the transient thermal impedance.

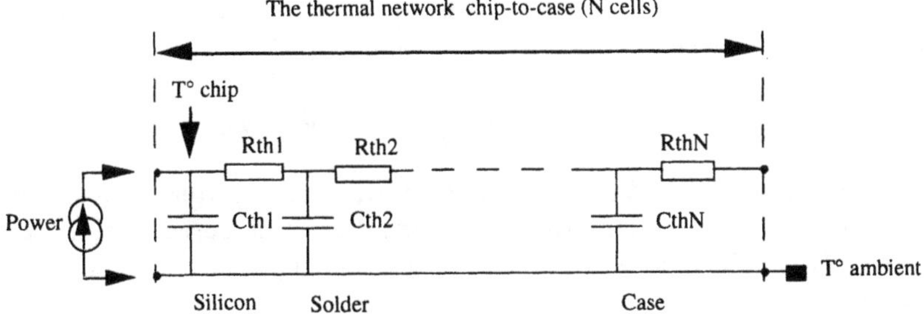

Figure 6:   The thermal network constitution (chip-to-case)

When the package geometry (chip area, material layer's thickness) is known [15], the thermal cells values can be calculated [16]. The lateral heat flow diffusion in the package is taken into account (Figure 7) by an appropriate diffusion angle θi between two layers.

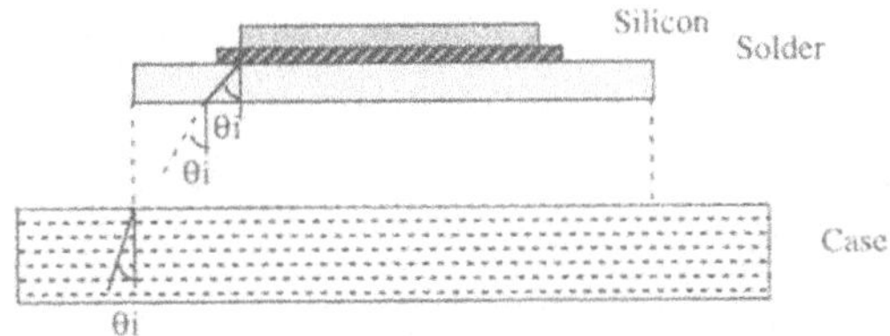

Figure 7:   Diffusion of the heat flow in "trunk of cone" taking into account the diffusion angle θi

For a material layer (i) :

$$Rth\ (i) = f\ \{area(i), thickness(i), Kth(i)\ , \theta(i)\} \qquad (2)$$

$$Cth\ (i) = g\ \{area(i), thickness(i), \rho(i)\ , Cp(i)\ , \theta(i)\} \qquad (3)$$

When the package geometry isn't known, the thermal cells value can nevertheless be determined using a numerical optimization (Figure 8). The single pulse response of the normalized transient thermal impedance (measurement [17] or manufacturer data sheet): r versus t constitutes the reference data.

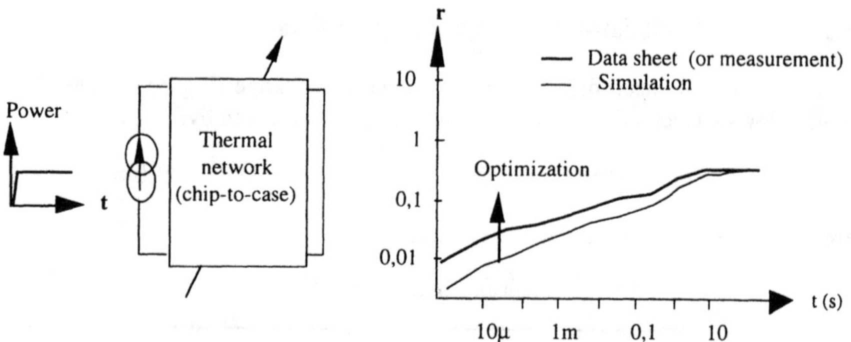

Figure 8:   Optimization of the thermal network with: r versus t

Figure 9 shows three curves of normalized transient thermal impedance: the experimental curve (manufacturer data sheet), the theoretical curve obtained with the calculated cells and the optimized curve. The last one gives a good result, because it rectifies the systematic error introduced by the first Rth-Cth cell, which doesn't adequately describe the heat propagation in the silicon chip.

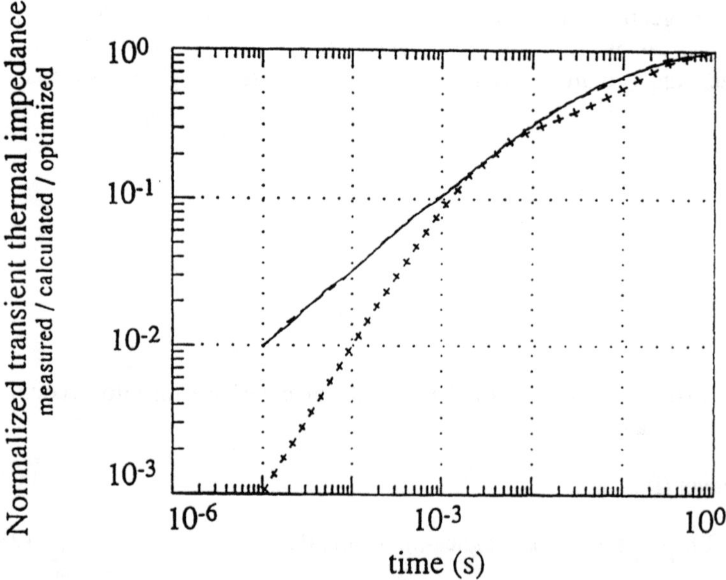

Figure 9:   Normalized transient thermal impedance - IRF 140 transistor
Measurement (solid line)
Simulation with calculated values (crossed line)
Simulation with optimized values (dashed line)

5.5.2. The Environment (Case-to-Ambient Air)

We must also consider the thermal environment of the transistor (Figure 10) [18]. To the conduction phenomenon (insulator, heatsink) can be joined the natural convection ($Rth_{case\text{-}ambient}$), the forced convection (ventilation) and the radiation, which can be modeled by additional $\{Rth(j)\}$ cells to the thermal chip-to-heatsink network. Formula appropriated to each phenomenon allow a calculation of them [18 - 22].

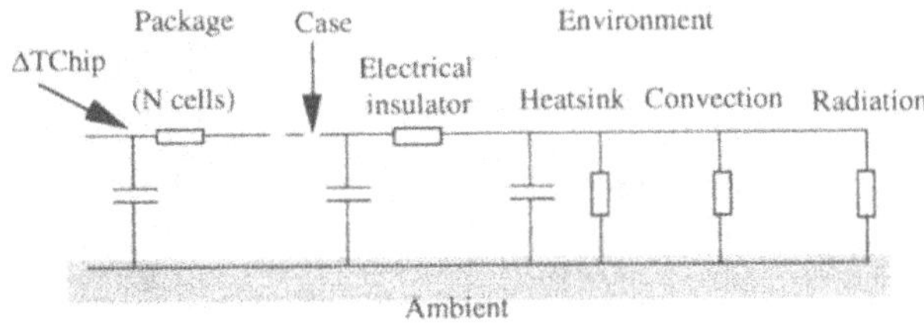

Figure 10:   The thermal network (chip-to-ambient)

5.5.3. The Mixed Thermal Model for the Power Bipolar Transistor

To improve the accuracy of the temperature calculation, this approach tries to take the temperature gradient existing in the bipolar structures into account. So, the previous Rsi-Csi cell, characterizing the silicon chip, is replaced by an analytical 2D solution, depending on the transistor geometry and the dissipated power profile [23,24] (Figure11.a).

The solution of the heat propagation differential equation takes the form of a series development and corresponds to the following geometrical assumptions and boundary conditions [25] :

- No temperature gradient along the z axis (uniform current density along the emitters).

- Uniform distribution of identical emitters along the x axis (x symetry).

- Dissipated power applied at the surface of the chip ($P = P(x, h/2)$).

- Constant thermal conductivity K0.

- No lateral heat loss ($\partial T/\partial x = 0$ for $x = \pm L/2$).

- Uniform temperature Tb at the bottom of the chip.

The hypothesis of a uniform temperature Tb at the bottom of the chip, and in the material layers of the case is based on the fact that the thermal conductivities of the different metallic layers of the case are much greater than the silicon's. It allows a more simple modelling of conduction phenomenon beyond the chip (solder-to-ambient) by the previously described Rth-Cth cells (Figure 11.b).

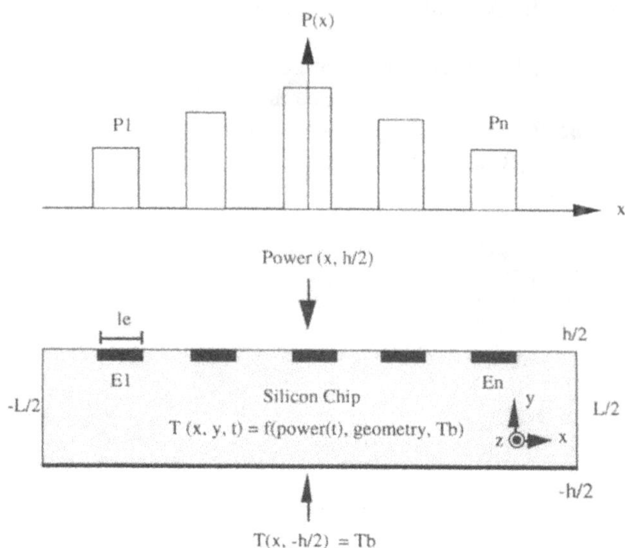

Figure 11. a: Boundary conditions and geometrical assumptions for the 2D thermal solution

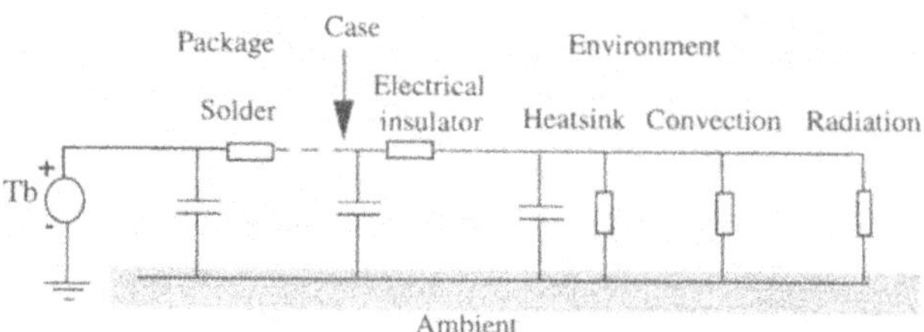

Figure 11. b: The thermal network (chip-to-ambient)

This 2D thermal solution for the silicon chip generates changes in the electrical model of the power bipolar transistor, that uses in this case the paralleling of identical electrical models (one per emitter finger), all having the same parameters, internal and external electrical nodes (common voltage drops), so that they only differ by their internal currents, charges and local temperatures. Each previously described electrical model is affected to the temperature located at the top center of its corresponding emitter finger $T(Ej)$ and dissipates the instantaneous power Pj ($ICj * Vce + IBj * Vbe$) under the emitter Ej, zero elsewhere.

The increase of the emitter resistors REi with temperature limits the current focusing under the central emitter during the simulation, so REi are equivalent to ballast resistors.

5.5.4. Stability and Convergence Time for the Complete Electrothermal Models

Since the thermal resistances and thermal capacitances used in the one-dimensional thermal model are linear elements, they do not introduce particular problems for the stability of the complete model and do not increase the simulation time for the same reason.

On the other hand, the use of the mixed thermal model for the power bipolar transistor increases the simulation time because in this case, the local temperatures T[Ej] need to be declared as state variables in the model and are considered as supplementary nodes. Each one depends unlinearly on the others and on the electrical variables, so the convergence of a simulation step includes both the electrical and the thermal convergences.

As an example of the simulation times required for the models with a SUN SPARK 2 station, the simulation of a DC curve (IC-VCE or ID-VDS) containing 40 points takes about 10 seconds using the one-dimensional thermal model and about 60 seconds using the power bipolar mixed thermal model.

5.6. EXPERIMENT - CHARACTERIZATION

The models parameters are extracted and optimized using the modeling software Ic_cap (with Veetest) and the Saber circuit simulator [26] (Figure 12).

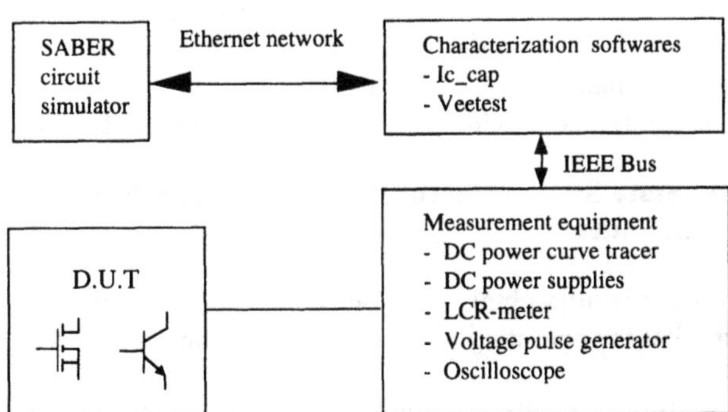

Figure 12:   Measurement benches

5.6.1. First Application: The VDMOS Transistor

The transistor used for this study is the IRF 140 in a TO-3 case which can be modeled as four cells (silicon, solder, copper and can).

The electrical network parameters (including electrothermal coefficients) are optimized using the IDS versus VDS characteristics (in pulsed mode) for different ambient temperatures. The dynamic parameters are measured in CV mode with a LCR-meter and optimized using the VDS(t) and VGS(t) curves in transient mode.

A fast solution to evaluate the electrothermal coefficients of the intrinsic resistances uses the Rds_{on} versus Temperature characteristic given by the manufacturer data sheet. The fit between the simulation of the variation law (4) and the experimental curve (Figure 13) allows a determinion of the two coefficients: c1 and c2 verifying the following equation :

$$R(T) = R(T0) \ [1 + c1 \ (T - T0) + c2 \ (T - T0)^2] \qquad (4)$$

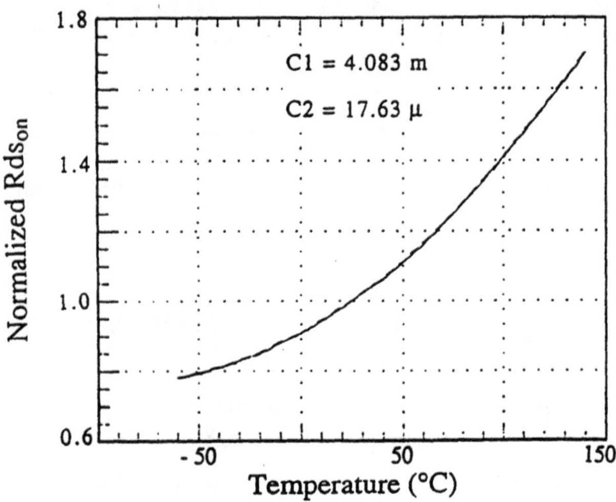

Figure 13:   Normalized Rds_{on} characteristic:
measurement (solid line) and simulation (dashed line)

5.6.1.1.   Steady-State Study (Measured and Simulated IDS .vs. VDS Characteristics)

Figure 14-a shows the results obtained in pulsed mode. The pulse widths (80 μs) are adequate to ensure that the operating point is established and to avoid self heating.

Figure 14-b presents the results obtained after thermal stabilization (5-10 seconds before each measurement point). The transistor is set on a water-cooled heatsink ($Rth_{heatsink}$ = 0.01 K/W), so the case temperature equals the water temperature. The current variations are dominated by simultaneous thermal effects (Figure 4). The simulated characteristic: ΔTchip versus VDS is also presented. ΔTchip increases with the electrical power dissipated. These results demonstrate that the model is able to accurately represent the electrical behavior of the VDMOS transistor with and without temperature effects.

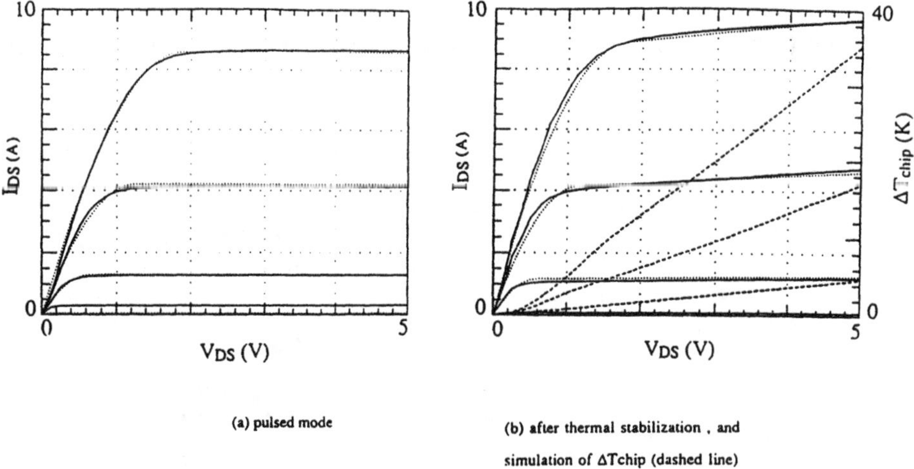

(a) pulsed mode   (b) after thermal stabilization , and
simulation of ΔTchip (dashed line)

Figure 14:   IDS - VDS step VGS characteristics,
Simulation (dotted line) and measurement (solid line)

5.6.1.2.   Transient Study

The transistor is used in transient mode on an R-L load (Figure 15). The simulated ΔTchip mainly increases during the commutations when the electrical power dissipated is the most important. The simulated and measured results are in good agreement and close the characterization stage of the VDMOS transistor.

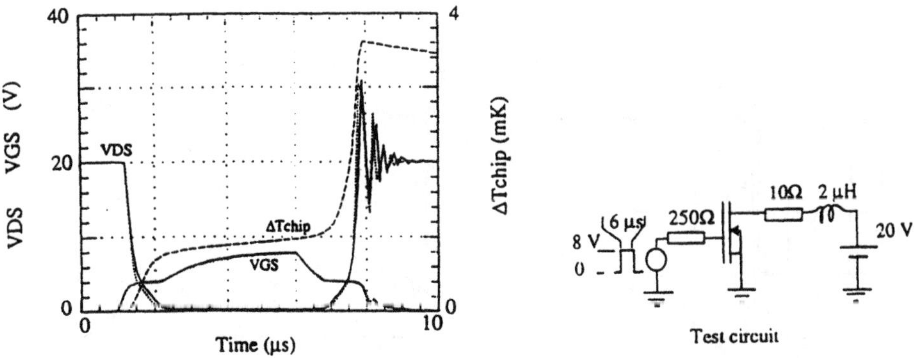

Figure 15: Simulation (dotted line) and measurement (solid line) of VDS(t), VGS(t)
and simulation of ΔTchip (t) (dashed line)

5.6.1.3. Validation Circuit

The circuit (Figure 16) has been characterized in transient mode for different input signal frequencies.

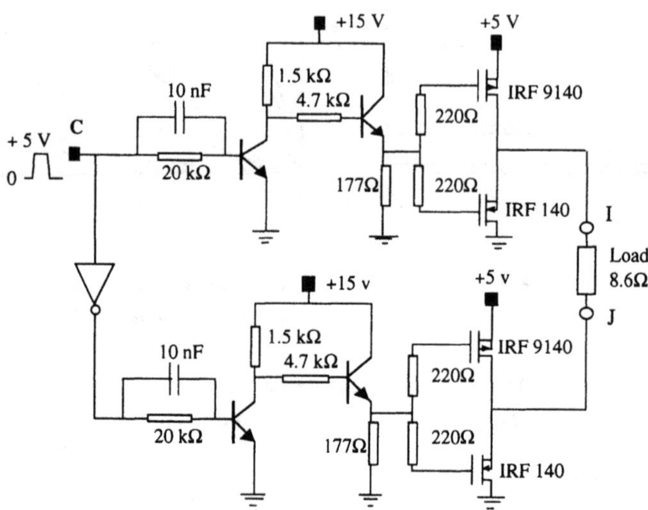

Figure 16: Test Circuit: The type of bipolar transistors is 2N2222

Figure 17 shows respectively the measured and simulated voltages on the C and I nodes at each frequency.

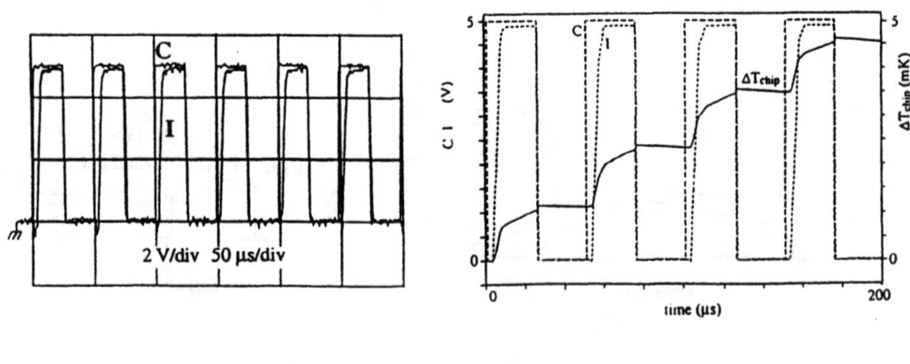

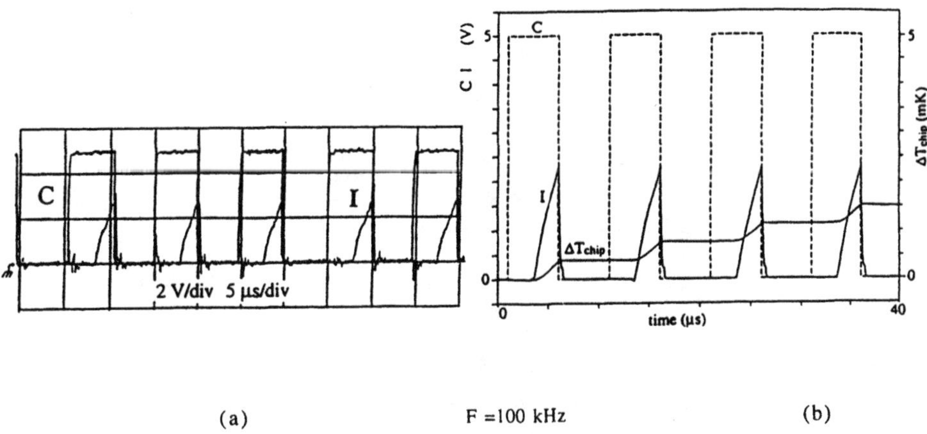

Figure 17:   Transient voltages on C and I nodes
and simulation of ΔTchip - Measurement (a) and simulation (b)

The accuracy of the signal transmitted to the load (turn-on and turn-off times, delay, max current) allows a precise determination of the cut-off frequency for this circuit. Note that in this case, the VDMOS transistors only dissipate power during the turn process and the temperature increase is negligible. However, the electrothermal interaction can become very useful for all the circuits where the transistors are submitted to much higher power dissipation [27].

5.6.1.4.   Thermal Stabilization Point (T.S.P) of a Power Transistor

The use of an interactive model for C.A.D can help a power circuit designer to determine the T.S.P. of the power transistors used in a given application. Nevertheless, the thermal time constants (> several seconds) are generally very important compared to the electrical time constants, so the circuit behavior must be simulated for several seconds to finally reach the T.S.P., consuming much simulation time.

One solution is to reduce the thermal network capacitances of the power transistors. The reduction must be enough to accelerate the thermal exchange between the chip and its environment and go more rapidly to the T.S.P., but without really modifying the electrical behavior. This allows a considerable reduction of the simulation time.

Another solution is to introduce initial thermal conditions that are near to the estimated final value for the chip temperature, before running a simulation with the right values of the thermal capacitances.

The first solution [27] is used within the example of a converter developed in a laboratory [28, 29] (Figure 18).

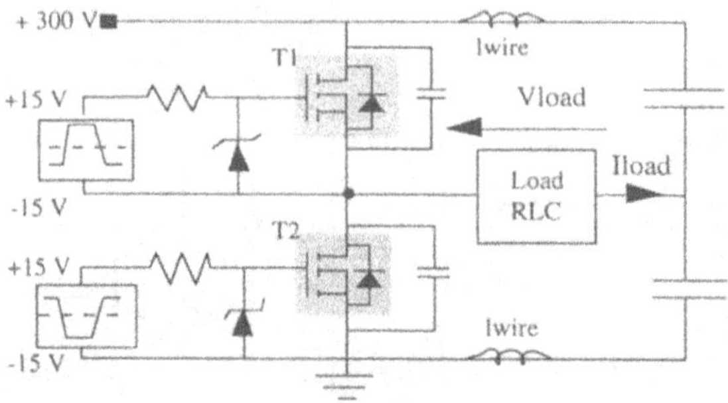

Figure 18:   Converter circuit using 15N40 VDMOS transistors
(Rth junction-case = 0.4 K/W)

Some tests with different commands fits have shown that the electrical signals on the load are quite unchanged, whereas the VDMOS thermal behavior is very different according to the cases.

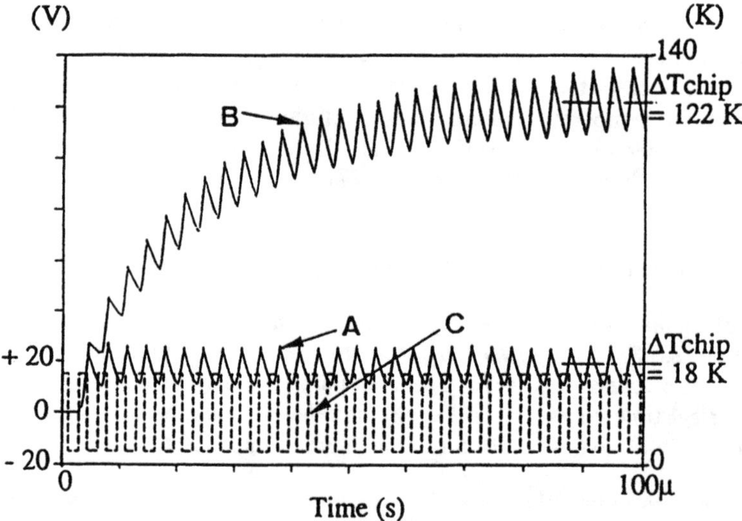

Fig 19:   Thermal stabilization point of a VDMOS (transistor T2) for two different
Rth j(unction)-a(mbient):
curve A: ΔTchip(t), for Rth j-a = 0.4 K/W
curve B: ΔTchip(t), for Rth j-a = 2.4 K/W
curve C: The corresponding VDMOS input command

Figure 19 shows two simulations with the same command inputs, for which the VDMOS have been set to two different heatsinks. The chip temperature of the VDMOS stabilizes to a mean value depending on the mean power they dissipate and on the total thermal resistance Rth junction-ambient.

5.6.2.   Second Application: The Power Bipolar Transistor

The Gummel-Poon transistor parameters are extracted and optimized with the standard procedure [30] using the Ic_cap-Saber interface [26]. The following procedure allows to extract and optimize the other transistor parameters.

5.6.2.1.   Steady-State Study in Pulsed Mode

The additional parameters for the epitaxial layer are extracted and optimized simultaneously on the two curves IC-VCE step VBE (Figure 20) and IC-VCE step IB (Figure 21). These first results allow a validation of the DC electrical model for the transistor. Repeating the operation for different ambient temperatures allows an extraction of the electrothermal coefficients.

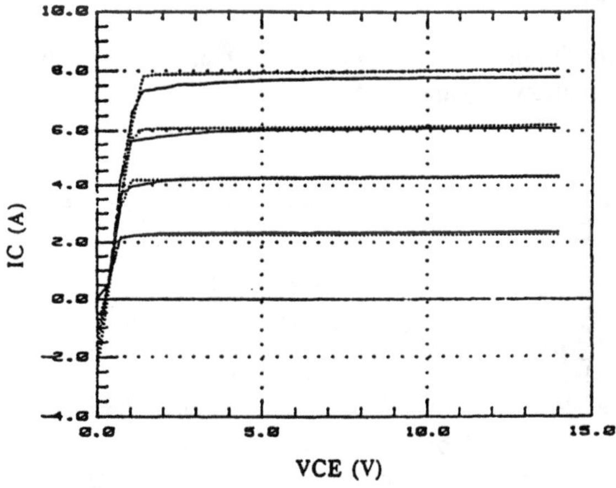

Figure 20:   IC-VCE step VBE network (VBE : 0.9 V-> 1.5 V, step 200 mV)
Pulsed mode measurement (solid line)
Simulation (dotted line)

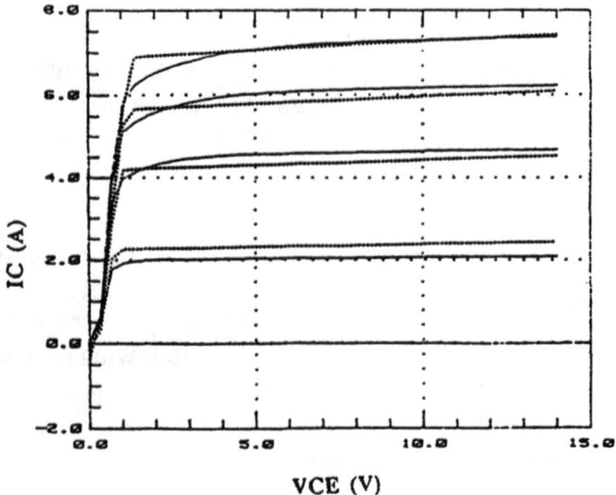

Figure 21:   IC-VCE step IB network (IB : 25mA -> 325mA, step 100 mA)
                Pulsed mode measurement (solid line)
                Simulation (dotted line)

5.6.2.2.   Steady-State Study after Thermal Stabilization (on a Water-Cooled Heatsink)

The current variations (Figure 22) are dominated by simultaneous thermal phenomena (Figure 5). This result allows a validation of the complete electrothermal model for the static behavior.

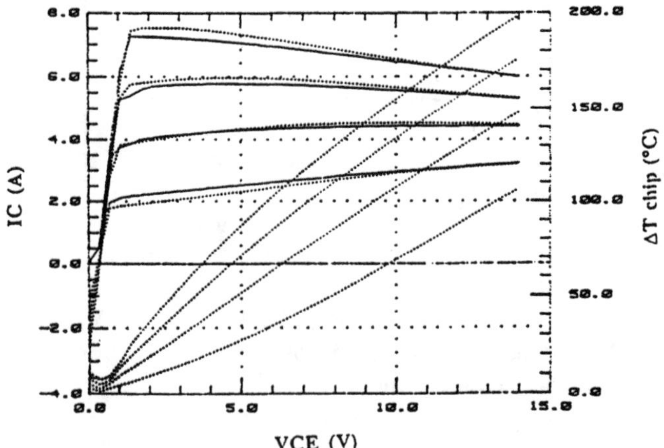

Figure 22:   IC-VCE step VBE network (VBE : 0.9 V->1.5 V, step 200 mV)
                Measurement after thermal stabilization (solid line)
                Simulation: IC and mean chip temperature (dotted line)

Figure 22 also shows, at each DC point, the evolution of the mean chip temperature obtained with a simple chip-to-ambient Rth-Cth thermal network. For a mean dissipated power of 65 Watts, the use of the chip 2D thermal model allows a first showing of temperature profiles in the chip for different depths from the surface to the bottom, and second, the associated power profile (Figure 23).

Note that in the same conditions, the simple Rth-Cth thermal network gives a mean chip temperature of 160 °C (Figure 22), that is near the mean chip temperature at y=0 (chip half depth).

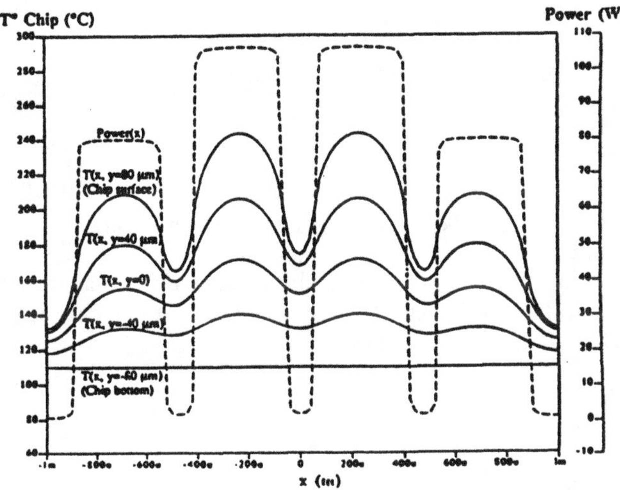

Figure 23:   Temperature and power profile
for P = 65 W ,VBE=1.5V, IC=6.5 A, VCE=10 V,
Chip geometry : 2mm*2mm*160µm (4 emitter fingers)

5.6.2.3.   Transient Simulation on Resistive Load

Figure 24 shows the simulated signals IC, VCE and IN (input signal). The delay before turn off (due to the epitaxial layer discharging process) shows the non quasi-static behavior of the stored charge Qepi.

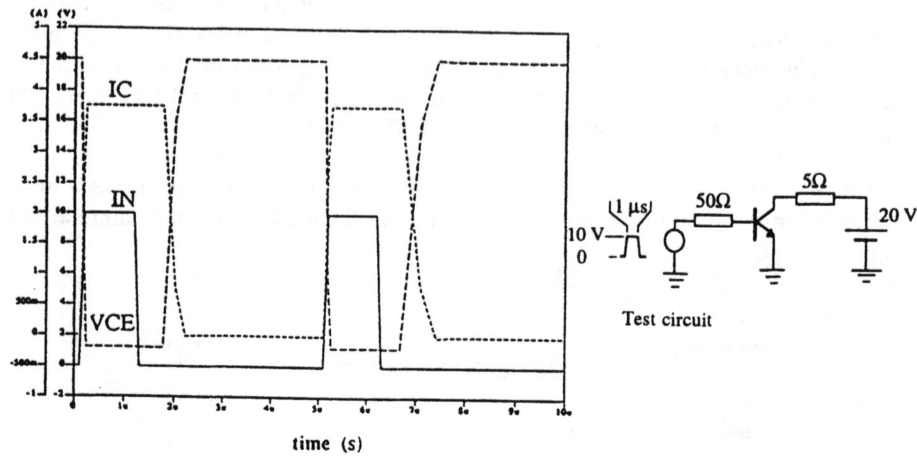

Figure 24: Output signals during switching

Figure 25 shows, in the same conditions, the simulated internal variables Qepi, the dynamic diffusion length λ, and ΔTchip associated to a chip-to ambient one-dimensional model.

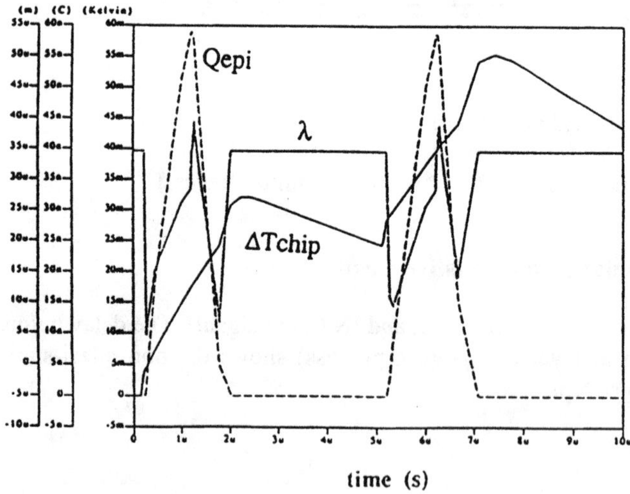

Figure 25: Internal quantities during switching

5.7.   CONCLUSION

We have developed accurate one-dimensional device models of power mosfet and bipolar transistors taking into account thermoelectric interactions and have shown that the influence of temperature on the electrical behavior of the transistors is well modeled, under static and transient conditions. However, the calculation of a mean chip temperature can sometimes introduce an important error on the chip temperature value, because of the existence of a temperature gradient in the chip for the bipolar technology. A temperature calculation using the heat propagation differential equations and an appropriate cutting-off for the silicon chip can been added to the model for more accuracy.

Appropriate extractions have been developed and provide accurate values for all parameters. The models can be used as performance analysis tools in power electronics, especially in circuits simulation, where they can help the designer to optimize the heat exchanges between the transistor and its environment. They allow a check to determine whether the application stays within thermal limits.

5.7.1.   Acknowledgment

The authors would like to thank Mr. Pierre Aloisi [15] Engineer to Motorola (Toulouse/France) for the geometrical data and others useful informations.

REFERENCES

[1]   K. Zamlynski and J. Ogrodsky: "Electro-thermal analysis of IC's", Proceeding of IEEE conference   ISCAS'94, 30 May - 2 June, 1994, London, Vol. 1, p. 17-20

[2]   E. Schurack, T. Ladzel and A. Gottwald: "SPICE- Simulation of nonlinear effects in field-Effect-transistors   caused by thermal power feedback", Proceeding of IEEE conference ISCAS'93, Chicago, Illinois, May 3- 6, 1993, p. 1116-1119.

[3]   A. R. Hefner and D. L. Blackburn: "Simulating the dynamic electrothermal behavior of power electronic circuits and systems", IEEE Transactions on power electronics, Vol. 8, N. 4, October 1993, p. 376-385

[4]   C. Lallement, R. Bouchakour and T. Maurel: "A VDMOS transistor model taking into account the thermal-electrical interactions", Proceeding of IEEE conference ISCAS'94, 30 May - 2 June, 1994, London, Vol. 1,   p. 317-320

[5]   C. Lallement and R. Bouchakour: "Temperature effect modeling on the electrical performance of power Mosfet", Proceeding of IEEE conference INTELEC' 93, France, September 27-30, 1993, Vol. 2, p. 235-241

[6]   C. E. Coordonnier, R. Maimouni, M. Belabadia, P. Rossel and H. Tranduc: "Power Mosfet Model Running on SPICE Program for Resistive and inductive Switching", 4th International Macroelectronics Conference, Proceeding of Electronica 88, Munchen, Germany, November 9-10, 1988, p. 58-71.

[7]   C. Xu and D. Schröder: "Modelling and simulation of Power Mosfets and Power diodes", PESC'88 Record, Kyoto, Japan, April 11-14, 1988, p. 76-83.

[8]   K. Shenai: "A Circuit Simulation Model for High- Frequency Power", IEEE Transactions on Power   Electronics, Vol. 6, N°3, July 1991, p. 539-547.

[9]   C. Xu, D. Schröder: "A power bipolar junction transistor model describing static and dynamic behavior" IEEE Transactions on Power Electronics, October 1992, Vol. 7, N 4, p. 734-740

[10]   H. K. Gummel, H. C. Poon: "An integral charge control model of bipolar transistors", Bell Syst. Techn Journal, Vol. 49, 1970, p. 827.

[11]   P. Antogretti, G. Massobrio: "Semiconductor device modeling with Spice", McGraw-Hill Book Company, 1988.

[12]   C. - P. Wan and B. J. Sheu: "Temperature dependence modeling for MOS VLSI circuit simulation", IEEE Transactions on Computer-Aided Design., October 1989, Vol. 8, N° 10, p. 1065-1073.

[13]   S. - W. Lee and R. C. Rennick: "A compact IGFET model - ASIM", IEEE Transactions on Computer-Aided Design , September 1988, Vol. 7, N° 9, p. 952-975

[14]   W.E Newell: "Transient Thermal analysis of solid state power devices making a dreaded process easy", IEEE PESC'75 record, 1975, p. 234-251

[15]   Geometrical Data of a manufacturer: Motorola.

[16]   C. E. Coordonnier: "Modèle SPICE permettant l'analyse en transitoire de la température des puces de puissance", Journée d'étude, Marseille (France), June 2- 3, 1989, p. 1-7.

[17]   D. L. Blackburn: "A review of thermal characterization of power transistors", IEEE 4th annual Semiconductor Thermal and Temperature Measurement Symposium, San Diego, CA, 1988, p. 1-7.

[18]   G. Owen : "Thermal management techniques keep semiconductors cool" Electronics, September 25, 1980, p. 135-142.

[19]   G. N. Ellison : "Thermal Computations for Electronic Equipment" New York : Van Nostrand Reinhold, 1984, p. 218

[20]   R. Siegel and J.R. Howell: "Thermal radiation heat transfer", Hemisphere, Washington, DC, 3rd ed., 1992.

[21]   RCA Designers "Handbook: Solid-State Power Circuits", Technical series SP-52, 1971, pp. 40-57

[22]   General electric SCR manual, 6st edition, 1979, p. 535

[23]   T. Maurel, R. Bouchakour and C. Lallement: "Modeling and simulation of the electrothermal behavior of the power bipolar transistor", 2nd European IC-CAP/HP4062 User Meeting, Horbourg-Wihr, October 10-11, 1994.

[24]   T. Maurel, R. Bouchakour and C. Lallement : "Modeling and simulation of the electrothermal behavior of the power bipolar transistor", Proceeding of IEEE conference PEDS'95, Singapore, 21 24 february, 1995.

[25]   William Liu and Burhan Bayraktaroglu: "Theoretical calculations of temperature and current profiles in multi-finger heterojunction bipolar transistors", Solid State Electronics. Vol 36, N° 2, 1993, p. 125-132

[26]   R. Bouchakour, C. Lallement and T. Maurel: "Thermal and electrical modeling and characterization of bipolar and power mosfet transistors with Ic_cap and Saber", Characterization Solutions Revue, Hewlett Packard, Winter/Spring 1994, p. 6-13.

[27]   C. Lallement, R. Bouchakour and T. Maurel: "One-dimensionnal analytical modeling of the VDMOS transistor taking into account the thermoelectrical interactions", Proceeding of IEEE conference MWSCS'94, Lafayette Hilton and Towers, Lafayette, Louisiana (USA), August 3-5, 1994.

[28]   K. al Haddad, Y. Cheron, H. Foch and V. Rajagopalan: "Static and small signal analysis of a series-resonant frequency", Can. Eng. J., Vol. 12, N° 4, 1987, p. 158-164.

[29]   J. P. Arches, N. Bonnet, F. Oms, D. Revel and J. Roux: "Transistors MOS haute tension: optimisation de la commutation", RGE (a French revue), N° 8, Septembre1989, p. 14-19.

[30]   I.E.Getreu: "Modeling the bipolar transistor", Elsevier, N.Y. 1978.

INDEX

Printed in the USA
CPSIA information can be obtained
at www.ICGtesting.com
LVHW010357161023
761183LV00007B/756